ENERGY POLICIES, POLITICS AND PRICES

ENERGY POLICY

PERSPECTIVES, CHALLENGES AND FUTURE DIRECTIONS

ENERGY POLICIES, POLITICS AND PRICES

Additional books in this series can be found on Nova's website under the Series tab.

Additional e-books in this series can be found on Nova's website under the e-Books tab.

ENERGY POLICIES, POLITICS AND PRICES

ENERGY POLICY

PERSPECTIVES, CHALLENGES AND FUTURE DIRECTIONS

K. J. SREEKANTH
EDITOR

nova
science publishers
New York

NOTICE TO THE READER

Library of Congress Cataloging-in-Publication Data

ISBN: 978-1-53613-744-6

Published by Nova Science Publishers, Inc. † New York

CONTENTS

PREFACE

Sustainable development is triggering a re-assessment of innovation and technological change in all fields, and energy is no exception. A key challenge of energy sustainability is to examine the range of credible potential pathways of combined social, environmental and technological systems under conditions of uncertainty, stagger, personal preferences and complication. Conventional energy resources – essentially fossil fuels – are becoming limited because of the swift increase in energy demand. This disparity in energy demand and supply has placed enormous coercion not only on consumer prices, but also on the natural world; this requires mankind to look for sustainable energy resources.

Sreekanth. K. J., PhD begins this book by first describing the energy efficiency and emission reduction characteristics of the road transportation sector in Chapter One. Chapter Two proposes the costs of renewable energy promotion and benefits through an analysis of the European case by Margarita Ortega Izquierdo and Pablo Del Río. Next, Chapter Three, by Jiang Yu and Zheng Fang, presents a review on residential electricity price policies in China. In Chapter Four, Fotouh Al-Ragom discusses the behavior change approach with a metric to promote and sustain energy efficiency. The Nigerian electricity market and its future is explained in Chapter Five by Karen Maguire and Kolawole Olaniyi. The institutionalization of the common gas market in the context of institutional evolution of the Eurasian economic union by Elena Shadrina is explained in Chapter Six.

Chapter 1 - Ensuring the circulation of passengers and goods between different areas is fundamental for the prosperity of any economy and for the welfare of people but at the same time, the complexity of transport's system produces relevant negative externalities. Transport sector accounted in 2015 for around 20% of global primary energy use and energy-related CO_2 and NO_x emissions mainly. This sector proved to play one of the main roles in energy issues from 2007 to 2030 for the increase of 97% in the world's primary oil use is expected. Given world's rapid economic growth, low price of old vehicles, and easy access to loans to finance new vehicle purchases, the vehicle fleet is expected to continue to grow in the coming years, and this will not be matched by similar growth in the capacity of road networks and transport facilities.

Transport sector's management practices are characterized by the unavailability of an integrated strategy. This chapter will review various approaches and technologies such as development of vehicle labelling for CO_2 emission and fuel efficiency and reduction in the speed limit of light commercial vehicles for enhanced energy efficiency. The study on different energy efficiency and emission reduction characteristics, which can be implemented throughout the world, through the identification of proper policies and plans to achieve a sustainable road transport system, is the major purpose of this work. In addition, to evaluate the effectiveness of different energy efficiency technology strategies for future system development pertaining to road transportation sector to reduce the GHG emissions is another feature of this study.

Chapter 2 - Major changes in behavior and technologies will be required in order to reach the ambitious decarbonisation goal of the Paris Agreement. Innovation and diffusion in low-carbon technologies will contribute to this goal. Among these technologies, renewable energy technologies have a main role to play. Indeed, it is widely acknowledged that electricity from renewable energy sources (RES-E) brings many environmental and socioeconomic benefits for society compared to conventional electricity. The comparatively lower social costs of these technologies (once their socioeconomic and environmental benefits are taken into account), their higher private costs, and the cost reductions that they experience with increased diffusion have been the reasons for their public promotion everywhere. However, such support has increased substantially in many countries around the world, raising the concerns of governments about the social acceptability and political feasibility of the policies. This has called into question the continuation of RES-E promotion. However, in so far as the socioeconomic and environmental benefits are high, RES-E support costs can be justified. The aim of this paper is to identify the main socioeconomic and environmental benefits of RES-E and to compare them with RES-E support costs, using a novel methodology developed by the authors in the past. The analysis is geographically circumscribed to the Member States of the European Union. It is found out that, indeed, the socioeconomic and environmental benefits in terms of CO_2 emissions reductions and fossil fuel savings are substantial in the analysed period (2010-2015). In monetary terms, the fossil fuel savings are more relevant than the CO_2 emissions savings. While the monetary calculation of the avoided CO_2 emissions has experienced an increasing trend over the period, the evolution of fossil fuel savings reached a maximum in 2013. The unitary support costs (€/MWh) increased over the period. The authors' results show that the total support costs have been above the socioeconomic and environmental benefits, although it should be taken into account that benefits and costs fall on different agents and, thus, they are not directly comparable.

Chapter 3 - This chapter reviews the history of residential electricity pricing in China in five stages. The first stage is a period during 1987-1997 when there was a power supply shortage, which led to an initial discussion on the feasibility of tiered electricity pricing. The second stage is from 1998 to 2003 when a surplus of electricity supply emerged and electricity pricing is used to promote the electricity consumption. Subsequently the authors focus discussion on the specifics and policy effects of the tiered electricity pricing implemented in three provinces Zhejiang, Fujian and Sichuan during 2004-2006. The fourth stage is a period during 2008-2010 when the expansion of the tiered electricity pricing policy is somewhat stagnated. Finally, the tiered electricity pricing is implemented nationwide since 2012, and the characteristics of the pricing policies across provinces and some policy effects are discussed in detail. The tiered pricing policies are still undergoing some adjustments and changes up till now.

Chapter 4 - The general trend in efficiency programs involves using technological and economical tools. These tools can support decision making specifically for program implementation justification (techno-economic validation). Once the programs are in place, the success of these programs lies in the proper adaptation and retained dedication of project stakeholders or end-users. Hence psychology and behavioral economics tools can be used to capture culture and psychology trends to devise comprehensive energy efficiency programs that can eliminate technologies adoption barriers. Such comprehensive programs will incorporate energy-focused behavior change interventions. This chapter reviews available energy-focused behavior change programs highlights behavior biases and gaps and describes a behaviorally informed design scheme that can be utilized as a tool for developing energy efficiency behavior change interventions.

Chapter 5 - There are two divergent views of the future of Nigeria's electricity supply, one focused on the lack of investment in grid-based electricity supply, which is leaving Nigerians without power. The second bursting with optimism that distributed renewable generation will solve electricity supply issues throughout sub-Saharan Africa. These two visions of the future of electricity in Nigeria provide a glimpse into the challenges of resolving complex infrastructure issues in developing regions. The optimistic view is based on Nigeria's renewable energy potential, while the pessimistic view has forefront in its mind the difficult challenges of finding sufficient capital investment to channel that renewable potential into powering the homes and businesses of Nigerians. In addition, historically Nigerians have had unreliable access to electricity at best and it is not clear that distributed generation will provide reliable access either. Developed nations today rely on grid-based power systems, in large part because of the redundancy and therefore reliability that they provide. Large scale adoption of a reliable distributed renewable system would leapfrog Nigeria and other sub-Saharan African countries into a new technological future, if it is possible.

Chapter 6 - The chapter addresses some of the aspects of the ongoing institutionalisation in the Eurasian Economic Union (EAEU). Having briefly overviewed the history of the EAEU formation, it describes the overall institutional environment of the EAEU and characterises the contemporary stage of the EAEU's economic integration depicting the impediments and the opportunities for the latter. The rest of the chapter is devoted to exploring the common gas market (CGM) recently endorsed by the institutional provisions of the EAEU. In doing so, it analyses contemporary gas relations within the EAEU with application of the most recent data published by the Eurasian Economic Commission, examines the CGM's model and its principal institutions as they are envisioned. This chapter discusses potential benefits and concerns of the economies engaging in the CGM, as well as speculates on possible developments involving third parties. It concretises those through the analysis of the most salient issues of the CGM's formation as they relate to gas price, pricing, competition and others. The chapter argues that despite long existing constraints for integration in the EAEU due to mainly the factor of the member states economies' structural similarity, institutionally compatible economies display their intention to progress beyond supranational institution-building to include sectoral initiatives, within which the studied example of the CGM falls. The chapter demonstrates that the CGM exploits the existing institutional complementarity among the respective national gas sectors and extends Russia's somewhat more advanced institutional arrangements to the projected common gas market.

In: Energy Policy
Editor: K. J. Sreekanth
ISBN: 978-1-53613-744-6

Chapter 1

ENERGY EFFICIENCY AND EMISSION REDUCTION CHARACTERISTICS OF THE ROAD TRANSPORTATION SECTOR

K. J. Sreekanth, PhD, S. Alkandari, S. Alosaimi and M. Al-Mansour
Energy and Building Research Center,
Kuwait Institute for scientific research, Kuwait

ABSTRACT

Ensuring the circulation of passengers and goods between different areas is fundamental for the prosperity of any economy and for the welfare of people but at the same time, the complexity of transport's system produces relevant negative externalities. Transport sector accounted in 2015 for around 20% of global primary energy use and energy-related CO_2 and NO_x emissions mainly. This sector proved to play one of the main roles in energy issues from 2007 to 2030 for the increase of 97% in the world's primary oil use is expected. Given world's rapid economic growth, low price of old vehicles, and easy access to loans to finance new vehicle purchases, the vehicle fleet is expected to continue to grow in the coming years, and this will not be matched by similar growth in the capacity of road networks and transport facilities.

Transport sector's management practices are characterized by the unavailability of an integrated strategy. This chapter will review various approaches and technologies such as development of vehicle labelling for CO_2 emission and fuel efficiency and reduction in the speed limit of light commercial vehicles for enhanced energy efficiency. The study on different energy efficiency and emission reduction characteristics, which can be implemented throughout the world, through the identification of proper policies and plans to achieve a sustainable road transport system, is the major purpose of this work. In addition, to evaluate the effectiveness of different energy efficiency technology strategies for future system development pertaining to road transportation sector to reduce the GHG emissions is another feature of this study.

Keywords: emission reduction, transportation, GHG emission, fuel consumption, policy planning

INTRODUCTION

All countries are now facing the challenge of reducing and mitigating carbon emissions, and no strange, that transportation sector has a great contribution in emitting carbon to the atmosphere. Transportation now accounts for about half of global oil consumption and one-quarter of global GHG emissions from fossil fuels [1]; Not only the world is facing the global warming and climate change challenges, there are also other challenges like, economic and GDP growth, social and lifestyle changes, the growth of fuel demand, urbanization, demographics related issues, new technology, and government policies. Many challenges are arising to indicate the importance of reviewing the transportation policies and look at its efficiency thoroughly.

The challenges which transport sector is facing are so many, and by decreasing the effect of one of these challenges, the other one can be adversely affected. For example, the problem must not be looked at from one angle neglecting other factors, but in a comprehensive way, where all factors are integrated and all these challenges must be studied, analyzed, and evaluated synchronously to have a sustainable and reliable transportation system.

Amongst road, aviation, and sea transport, the road transport accounts for more than three-quarters of total transport demand. While many researchers and experts predicted that, the world's motor vehicle population would double in the next twenty years, for that the highlight of this chapter will be on the road transport. The United Nations projects the world's population to grow by 0.9% per year on average [2] Social factors and people's lifestyles have a very critical role in the transportation sector, and they vary from region to another. The increase in public awareness for example, and changing the driver behaviour will -of course- aid in the process of improving transportation sector.

Countries are now implementing some of the solutions to lower the carbon emissions, also to decrease fossil fuels and increase the use of renewable; one of the implementations is using biofuels. However, the biofuels industry is still developing, new and more advanced biofuels are discovered. With this great interest and effort on biofuels, still, the effect from a global perspective is not that promising. Statistics show that in 2010, renewable accounted for just 2.5% of total transport and 3.3% of road transport energy consumption [3]. Furthermore, Researchers and experts are debating on the sustainability and the side effects of biofuels like the effect on crops, food, agriculture, water, etc. other difficulties that face the biofuels implementation more widely are the high costs.

Electric vehicles and hybrid electric vehicles are also being manufactured by big and famous companies. Electrification of transport is helping with mitigating local pollutant but on the other hand, will result in high electric demand from power plants, which also depend and rely in so many countries mainly on fossil fuel. However, with the countries that are cutting their dependence on fossil fuels and are building renewable and green power stations, transport electrification will solve more than local but global pollutant and GHG emissions too.

TRANSPORTATION FLEET PATTERN IN KUWAIT

In Kuwait, the main transportation fleet pattern is private, while the minority party is public which is forsaken in general. Most of the "car owners" prefer to use their own vehicles for transportation purpose (citizens and non-citizens), and that includes the buses as well and not only the cars. The number of private buses registered in Kuwait in 2014 was 27,000, while on the other hand the number of the public buses registered w 3,000. The total number of vehicles used in Kuwait is growing every year due to the growth of the population, and, the number of vehicles registered in Kuwait that are in use in 2012 was 1.6 million, while it became 1.7 million in 2013.

Kuwait is placed on the top of the pyramids for the highest number of car owners in the Arab countries and is placed sixth globally, while the US is placed third according to the AUTOSTAT. In Kuwait, it was found that for every 723 vehicles there are 100,000 people. The increased number of car owners is due to multiple reasons that the public is mostly unaware of. The first reason is the somehow high tax-free income for the Kuwaiti citizens, which is also supported by the government with different allowances. Secondly, Kuwait is one of the leading countries around the world on producing oil and a member of the Organization of the Petroleum Exporting Countries (OPEC), and the availability of oil led the government of Kuwait to offer the gas with the cheapest possible price. Also, carpooling is not a very common thing in Kuwait.

Avoiding the switch of using public transportation (especially buses) is a very strong reason on why the congestion occurs during peak hours in Kuwait and is also one of the main reasons, if not the most reasonable, on increasing the emissions and the air pollution in Kuwait and the rest of the world. The reasons on why the car owners prefer to use their private vehicles and avoid the public transportation vary from one family to another and from one person to another, and it is very difficult to control these reasons as it is very difficult to set a law that will check the number of vehicles an individual should own and even harder to force them to use the public transportation especially that the public transportation in Kuwait is that of a low quality. A study by Koushki was made to identify specifically why parents refuse to use buses as the main transportation vehicles for their children to go to school and prefer to use their own vehicles and estimated the likely energy and air pollution savings for when the switch to using public transportation actually occurred [4]. On the side effects of congestion and air pollution, some studies in the UK shows that air pollution is 10 times higher than car accidents on the number of deaths per year, and the congestion and fuel consumption costs around $40 billion per year [4].

Moreover, the public transportation in Kuwait started promoting in 1962 with the beginning of Kuwait Public Transportation Company (KPTC), and now only three non-governmental public transportation companies exist in Kuwait. These public transportation companies do not follow any specific standardization of any bus types, and only focuses their buses on the areas with the higher demands of public transportation (buses). With that being said, the demand for buses nowadays is reduced due to many reasons, mentioned previously, like the affordable vehicles and petrol cost. Also, the quality and bad timing of this mode of transportation (buses) are very efficient reasons on why the public is avoiding the use of buses nowadays (Table 1).

Table 1. Response-Ranked Reasons for Not Using School Bus [4]

Reason	Frequency	Percent	Rank
Busing is very slow	229	18.6	1
Busing not available	212	17.2	2
Busing is uncomfortable/inconvenient	211	17.1	3
Busing has high accident rate	122	9.9	4
Busing is socially unacceptable	113	9.2	5
No one to accompany child	87	7.1	6
School is close to home	62	7.0	7
Bus drive is not married	61	5.0	8
The fee is high	60	4.7	9
Other reason	75	6.1	-

(Source: http://ascelibrary.org)

Furthermore, the transportation pattern in Kuwait is highly unlikely to change and is getting harder to monitor due to the increase in population and private vehicles owners. The number of vehicles on the road is very condensed, the drivers violate the road traffic rules, no regulations on the number of vehicles an individual can own, and the people are not yet willing to switch to the use of public transportation due to its low quality, bad timing and terrible services.

ENERGY CONSUMPTION AND EMISSION IN KUWAIT

Kuwait is one of the highest per capita consumption countries around the world due to many reasons, one of which, if not the main reason, is the high demand on electricity in the summer due to the extremely high temperatures. During summer times temperatures in Kuwait could reach mid-fifties and up (in C°), and the usage of A/Cs gets very high, which causes an increase of energy consumption and at the same time will increase the air emissions (due to R22 Refrigerant). The increased number of vehicles in Kuwait will add to the high emissions rate due to A/C usage.

The relationship between the car emissions and how much it consumes is directly proportional because the more the car consumes fuel the more it will emit. For instance, according to the WLTP FACTS, a diesel car emitting 95g CO_2 per kilometre consumes around 3.7 litres of fuel per 100km, while a petrol type car consumes around 4 liters/100km for the same CO_2 emissions [5]. The topic on "Global Warming" became more interesting to the world recently, as in the past several years before the development of high technology the main concern on manufacturers was to just make machines, vehicles, industries, etc. and not to care about the environment. Nowadays the awareness of the global warming effects has increased too which organizations and governments have set rules and regulations towards saving the environment and reducing the emissions. Air pollution measures have prompted changes in engine design, emission control devices, reformulated gasoline options, and vehicle testing among others. Car emissions levels depend on factors such as vehicle operating conditions, fuel characteristics, shown in Figures 1 and 2, and ambient conditions [6].

The brand and the age of the vehicle will definitely affect the number of emissions the vehicle will produce due to the type of fuel the vehicle use and whether it consumes less (environment-friendly). Also, the age plays an important rule on the number of emissions the vehicle will produce because relatively modern vehicles are designed to consume less fuel comparing to the older vehicle brands, which it automatically means that the newer brands will have fewer emissions compared to the older brands.

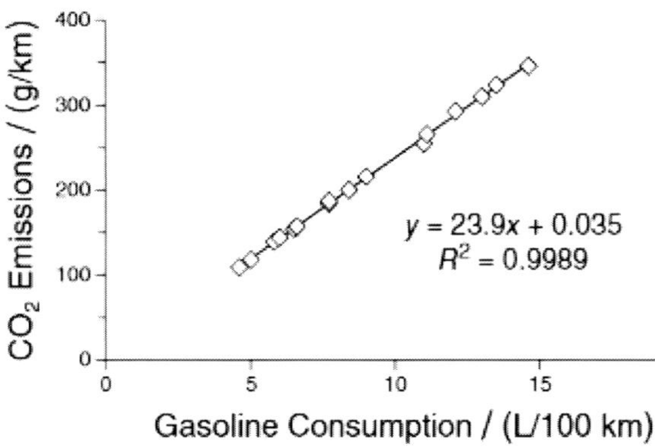

(Source: http://pubs.acs.org)

Figure 1. CO_2 emissions versus gasoline consumption [6].

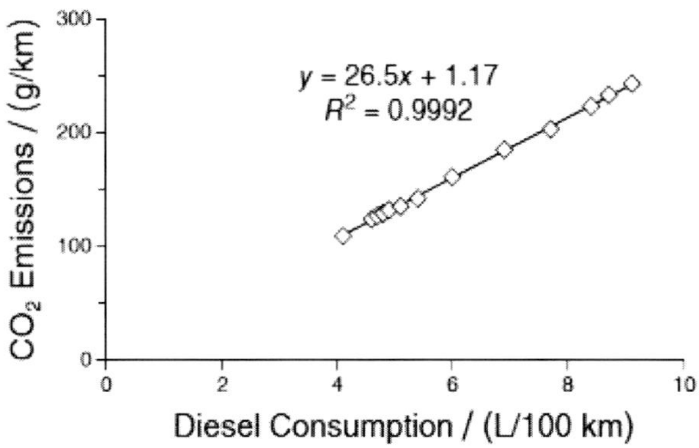

(Source: http://pubs.acs.org)

Figure 2. CO_2 emissions versus diesel consumption [6].

Table 2. Car Models Using Gasoline [6]

Consumption/(L/100 km)	Emission (g/km)	Car Model
11.1	266	Audi Carbriolet
9.0	216	Audi Sportback
12.1	293	BMW X3
7.7	187	BMW miniCooper
13.0	310	Chevrolet Corvette
14.6	346	Chevrolet Tahoe
8.4	200	Chrysler Sebring
11.0	255	Chrysler 300M
5.0	118	Daihatsu Sirion
6.6	157	Lancia Musa
5.8	139	Opel Agila
4.6	109	Peugeot 107
13.5	324	Quattro RS4
6.5	154	Skoda Fabia
6.0	144	Volkswagen Polo
7.7	184	Volvo S40

(Source: http://pubs.acs.org)

Table 3. Car Models Using Diesel [6]

Consumption/(L/100 km)	Emission (g/km)	Car Model
8.4	223	Audi Carbriolet
6.0	161	Audi Sportback
9.1	243	BMW X3
4.8	129	BMW mini
8.7	233	Chevrolet Captiva
7.7	203	Chrysler Voyager
6.9	185	Chrysler PT Cruiser
5.1	135	Lancia Musa
4.9	132	Opel Agila
4.1	109	Peugeot 107
4.7	127	Skoda Fabia
4.6	124	Volkswagen Polo
5.4	142	Volvo S40

(Source: http://pubs.acs.org)

Tables 2 and 3 provide data from various car (vehicles) models on two types of fuel that each brand could use, how much each car emits, and also how much each car consumes. It is known that the Diesel fuel vehicles are about 40% efficient and Gasoline fuel vehicles is about 35% efficient, which indicates that Diesel fuel vehicles are more efficient than Gasoline fuel vehicles. According to the data from Tables 2 and 3 the consumption and the emissions of a car using Diesel fuel are relatively less than it using Gasoline fuel (neglecting what the cost would be), which agrees with the previous statement of the efficiency of both Diesel and Gasoline fuels.

The idea of switching to hybrid vehicles/electric cars has been suggested or mentioned to the world for reduced polluting emissions and improved performance conditions, however; the high/expensive cost for each electric unit is one of the main reasons why people are not leaning towards the idea of "electric cars". Secondly, and most importantly, the oil and refinery industries will get highly affected by this switch as their income will be reduced. Also, so far, the hybrid vehicles are not made to withstand all the conditions, especially rough conditions like those in Kuwait, GCC, Alaska, etc.

The conditions in Kuwait are different than many other countries. The idea of using of electric vehicles might not be welcomed to the public yet. The reasons behind that are the availability of cheap oil and cheap government supported electricity, oil revenues are almost the only income source for the government of Kuwait, and, in general, the electric cars companies have not yet met all the requirements and needs of all the possible customers. A study was made in 2016 about the possible use of hybrid and electric cars in Kuwait. In the study, the comparison between three different modes that a vehicle could be used in Kuwaiti conditions and circumstances were made. The first type was purely mechanical (gasoline/diesel) engines, where he mentioned the cost and the consumption of the vehicle. In the study, it was found that the energy cost when using a mechanical-mode-engines is 3.04 fils/km by assuming the vehicle is running through a constant speed of 50 km/h the specific fuel consumption was 7.6 lit/100 km (the cost of fuel in Kuwait in 2016 was 65 fils/lit). In the other hand, for electrical-mode-engines, shown in Figure 3, the electrical energy cost came out to be 22.35% of the cost of petrol (the cost of electrical energy in Kuwait in 2016 was 2 fils/kWh). The last mode was the hybrid-mode-engines which, shown in Figure 3, unlike the electric-mode-engines, is an engine type that runs some and not all of its power by gasoline. The vehicles were tested under specific conditions where they took the minimal fuel consumption ranges between 2300-4200 RPM. in the end, it appears that the percentage of savings were between the range of 3.5 to 35.8%, which indicates that there is a possibility of making the idea of using hybrid vehicles in Kuwait [7].

EMISSIONS

Kuwait is an industrial country and one of the world leaders in producing oil with 14 oil fields, 21 gathering centres, and 3 large oil refineries with a huge capacity. In addition to oil production, Kuwait's main source of energy is Power stations, and the main source of fresh water is the desalination plants. Furthermore, the total number of vehicles per head is relatively large to the size and population of Kuwait where for every 723 vehicles there are 100,000 people. The weather in Kuwait faces extreme conditions throughout the year where the temperature at the summer could reach up to 55°C including dust storms, humidity, and arid periods. Due to the previously mentioned conditions, the demand for electricity is very high, and that led the consumption of power per head for the Kuwaiti individual to be the highest in the world. According to a research study on the emissions and air pollution in Kuwait, the high emissions that were caused by the previously mentioned points are highly effective on human health [8]. It is also mentioned that the percentage of SO_2, CO and NO_X emitted daily are very high and showed, in Figure 3, the percentage of each produced by power and desalination plants, motor vehicles, refinery fields, and oil fields.

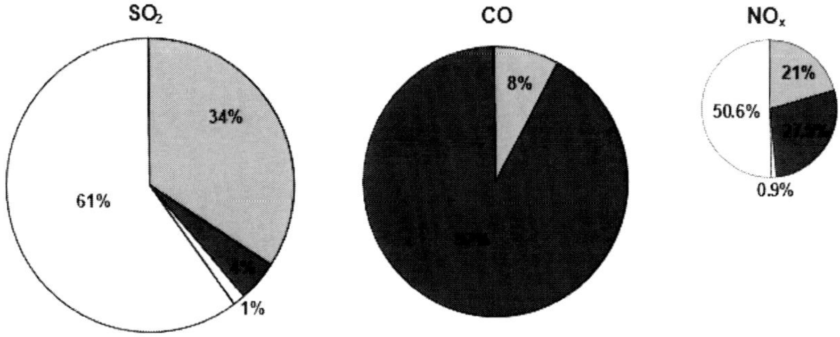

Source		Oil field	Motor vehicles	Oil refineries	Power plants	Total
Pollutant level, t/h	SO_2	9.133	1.2	0.382	16.095	26.81
	CO	1.712	19.69	---	---	21.402
	NO_x	1	1.31	0.047	2.414	4.771

(Source: http://www.tandfonline.com)

Figure 3. Source contribution to hourly emissions of SO_2, CO and NO_X [8].

Table 4. Emissions rate from motor vehicles, power plants, oil fields, and oil refinery fields in Kuwait [8]

	Value	Unit
Stack temperature	50	°C
Molecular weight of stack gas	29	kg/kmole
Outlet velocity of stack gas	15	m/s
Percentage of nitrogen in outlet stack gas	80	
Refineries		
Specific CO_2 emissions	0.01077696	kg/barrel
Specific SO_2 emissions	2.01908652	kg/barrel
Specific NO_x emissions	130	ppm
Total production capacity	850,000	barrel/day
Hourly emissions of SO_2	381.68	kg/hr
Hourly emissions of CO_2	71,509.31	kg/hr
Total hourly mass flow rate emissions	359,454.99	kg/hr
Total hourly volumetric flow rate emissions	21,901.59	m^3/hr

Table 4. (Continued)

	Value	Unit
Stack diameter	1.23	m
Hourly emissions of NO_x	46.73	kg/hr
Motor vehicles		
Specific CO emissions	12.98	gm/km
Specific NO_x emissions	0.86	gm/km
Annual travelled distance	20,000	kg/yr
Annual CO emissions per car	259.73	kg/car/yr
Annual NO_x emissions per car	17.23	kg/car/yr
Number of cars per 1000 inhabitant in Kuwait	332	
Kuwait population	2.00E+06	
Total number of cars in Kuwait	6.64E+05	
Hourly CO emissions	19,687.53	kg/hr
Hourly NO_x emissions	1309.36	kg/hr
Oil fields		
Hourly SO_2 emissions	7420.1	kg/hr
Hourly CO emissions	1712.33	kg/hr
Hourly NO_x emissions	1141.55	kg/hr
Power plants		
Specific emissions rate of SO_2	1	kg/MWh
Specific emissions rate of NO_x	0.15	kg/MWh
Specific emissions rate of CO_2	600	kg/MWh
Power capacity of Subiya power plant	4173	MW
Hourly SO_2 emissions	4173	kg/hr
Hourly NO_x emissions	625.95	kg/hr
Hourly emissions of CO_2	2,503,800	kg/hr
Total hourly mass flow rate emissions	12,542,994.75	kg/hr
Total hourly volumetric flow rate emissions	764,244.75	m^3/hr
Stack diameter	4.2	m

(Source: http://www.tandfonline.com)

Table 4 shows the value/amount of different variables that will affect the emissions rates for refineries, motor vehicles, oil fields, and power plants.

Furthermore, regarding the emissions in Kuwait, the Carbon Dioxide Information Analysis Center (CDIAC) experimented the total number of emissions in Kuwait and how much it increased from the year 1946 to 2008. The centre collected data regarding the emissions in Kuwait from following parameters: Gas Fuels, Liquid Fuels, Solid Fuels, Gas Flaring, Cement Production and Bunker Fuels. The percentage of Flaring, Cement and Solid Fuels either decreased or had no significant effect on the total percentage of Fossil-Fuel emissions. In the other hand, the Liquid Fuels and the Gases Fuels increased throughout the years and had a bigger role in increasing the total Fossil-Fuel emissions percentage, shown in Figure 5 [9].

From Figure 4, it can be seen that the data starts to have more significant change from1970 until 2008, while the data collected starts from 1946. The reason is that the data from the 1940s were on the pre-industrial era of Kuwait, however; in the post-industrial era, the data starts to grow and show more significant change on the total percentage of Fossil-Fuel Emissions [9].

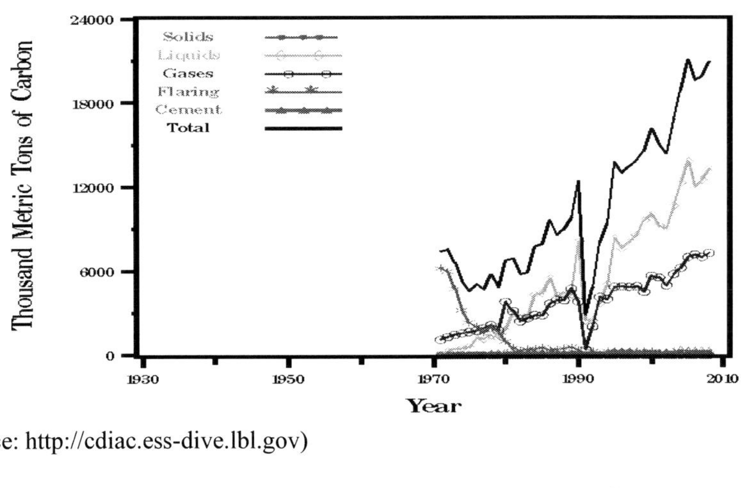

(Source: http://cdiac.ess-dive.lbl.gov)

Figure 4. Thousand tons of CO_2 emissions in Kuwait from 1970 until 2008 [9].

TRANSPORTATION POLICIES AND POSSIBLE SOLUTIONS

In a study pertaining to Kuwait, "Air Pollution and Road Traffic in Kuwait", it was presented that since Kuwait is having one of the highest Gross Domestic Product (GDP) in the world and the low fuel prices, it lead to the increase rate for ownership of motorized vehicles throughout the past years [10]. The transportation sector in Kuwait is a huge consumer of energy, wherein 1995 it was found that more than 24% of oil products in Kuwait were consumed by the transport sector, which is high compared to the other GCC countries, shown in Table 5. Table 5 shows that the amount of fuel consumed daily in Kuwait comes in second place right after Saudi Arabia (KSA), the largest country in the GCC area, where Kuwait is smaller than most of the GCC countries.

In the previous study, it was found that the NO_X, CO, SO_2 and HC emissions were much higher in the commercial and urban areas than the non-urban areas due to the high concentration of vehicle owners to live on those areas.

As mentioned before, Table 5 shows the data, collected from 1999 to 2001, on the total amount of gasoline consumed daily (1000 B/day) among the GCC countries, and it was found that Kuwait comes in second place only after Saudi Arabia even though UAE is much larger in Kuwait (size-wise).

Table 5. Gasoline consumption and percentage share of gasoline in total oil consumption in GCC countries [10]

Country	Gasoline Consumption (1000 Barrels/Day)			Percent of Gasoline of Total of Oil		
	1999	2000	2001	1999	2000	2001
United Arab Emirates	34.3	32.6	32.7	11.9	11.1	11.0
Bahrain	7.8	8.4	8.8	39.0	38.2	38.3
Saudi Arabia	219.6	223.7	224.3	20.1	19.2	19.1
Qatar	10.3	10.9	10.8	46.8	47.4	47.0
Kuwait	39.4	39.3	39.3	24.0	24.3	24.1

(Source: https://www.witpress.com/elibrary/wit-transactions-on-the-built-environment/75/12202)

Reasonable solution would be to set a policy on the amount of vehicles an individual could own (even though it contradict with their rights/freedom), increase the awareness of green-energy and the use of environmentally friendly cars, and increase the cost of fuel (Gasoline/Diesel) to the limit the use of the vehicles. Other solution would be to set a policy where each car-owner whose car is causing damage to the environment by releasing emissions higher than the normal range of the brand itself would be fined a specific amount of money (like Colorado, USA). These policies might have a positive effect on solving the transportation/emissions problem in Kuwait, however; the policies cannot override the rights and freedom of each individual on what they choose or how they live their lives (Figure 5).

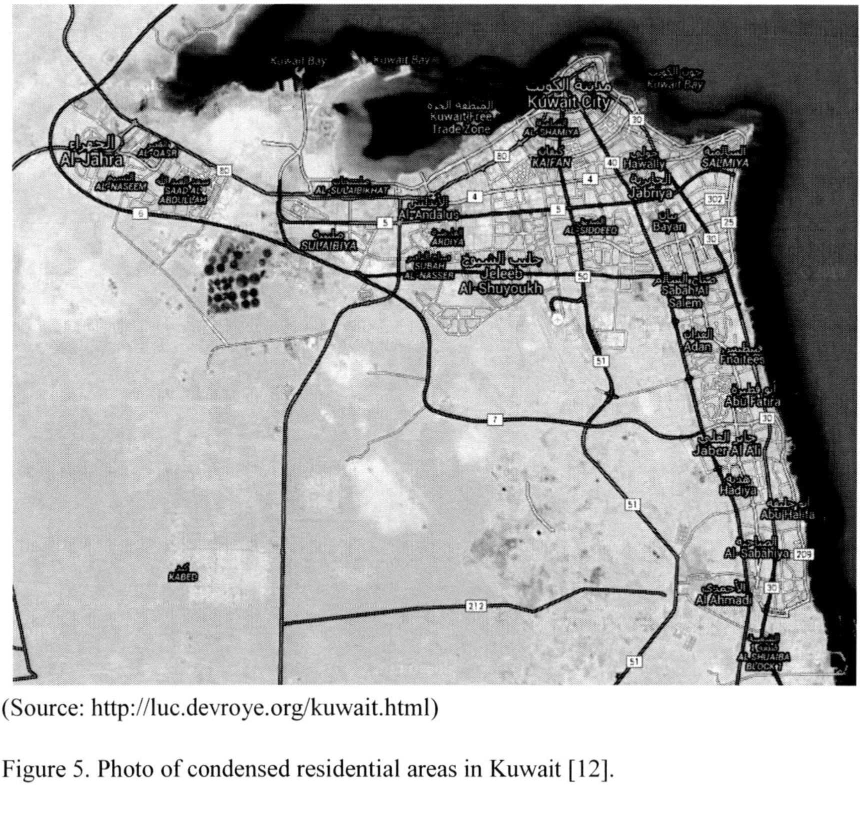

(Source: http://luc.devroye.org/kuwait.html)

Figure 5. Photo of condensed residential areas in Kuwait [12].

Furthermore, the reason why Colorado set some strict rules on the vehicles' emissions is that some cars cause more damage to the environment (atmosphere) than others. Another study on light-duty motor vehicle exhaust particulate matter measurements in The Denver, Colorado, Area," found that the average of emissions on newer vehicles w very low comparing to the average of older vehicles where the average increases accordingly to the age of the vehicle [11].

Switching to Green Energy in Road Transportation

The transportation sector is one of the largest contributors to greenhouse gas emissions and air pollution. Cars, trucks, aircraft, and railroads are contributing to transportation sector emissions, the reason behind that is consumption of oil and all types of transportation are dependent on fossil fuels. These emissions have a negative impact on health and environment. Nitrogen oxide and hydrocarbons are omitted from vehicles exhaust and affect air quality. These emissions in the air can cause respiratory difficulties for people [13].

Nowadays many countries in the world are looking forward to greening transportation sector by developing different clean alternative energy sources to eliminate emissions and solve climate change problems. These alternatives should not emit toxic gases that could affect the environment and human health [14].

There are many explanations for green transportation and one of them is "A truly green infrastructure is one that residents enjoy, provides durable and inexpensive mobility and addresses the underlying conditions from which our energy crises arise". Many organizations and researchers discuss in their papers the tendency of using Renewable Energy and Electricity as clean alternative energy sources in transportation instead of burning the fuel. There are many means of transport, which are environmentally friendly such as Bicycle, Electric bike, Electric vehicles, Green trains, Electric motorcycles, multiple occupant vehicles, and Hybrid cars. These means have a wide range of benefit in environmental and health side; also, they have a lot of benefit in economic and individual budgets.

Summary and Policy Implications

Undoubtedly, a lot of care should be taken when elucidating the results and then drawing policy implications from not only this transportation energy system modelling but any energy systems modelling attempt. Any developed model of the global energy and the corresponding transportation system has its own limitations. Seeing as the enormous complexity of these systems, generalizations have to be made, both as regards technological alternatives and pathways, regional differences and behavioural aspects. Changes in the energy modelling constitution and the influencing parameters could certainly change the results. Any model results should underline cost-effective pathways to low GHG emission prospects. On the other hand, this does not mean that governments should adopt policies that make sure that these particular energy strategy futures turn up. Significantly, our view is that a collection of policies and strategies should be implemented in the country to keep the transportation sector on track on the road to a low carbon dioxide GHG emission reduction objective.

- **Carbon Tax**: The most important strategy is the introduction and constant growth in the cost of emitting CO_2. This cost could be set up by either a carbon tax or a cap and trade system. Obviously, the transportation sector has to pay the carbon fee.
- If state governments want to ensure liquid biofuels in the transportation sector, then individual directives must be introduced in that sector, either in the form of higher carbon taxes or compulsory introduction of biofuels.
- H**ydrogen future and other advanced technologies**: Innovative technologies and subsequent technological systems are not developed at once. The introduction of carbon taxes/emission trading systems today will encourage the expansion of impending

cutthroat technologies such as biomass and wind but will do trivial to augment R&D on and manufacturing of more advanced technologies such as solar PV, fuel cells, etc. The development of such technologies in full swing takes several decades. Concerning the transportation sector, policies are necessary for the development of hydrogen vehicles and infrastructure. This will have to have an effect over the next couple of years in order to make competition on mass markets successful.

- All forms of energy subsidies should be made reliant upon the following circumstances: (a) they should set in motion a process of self - maintained growth, driven by dynamic learning and scale effects. (b) The subsidized technology should have a good possibility for capturing future markets. Subsidies should then decrease over time and the scale of the forte markets made accessible should be in proportion to the needs of the new technology.
- **Techno-diversity**: This aspect should be promoted. Even if many studies strongly suggest that hydrogen will be the key fuel for the future, governments and companies should also continue research efforts, including minor market experiments, on a number of technologies for alternative fuels production and use, e.g., woody biomass, etc.

In addition, security of supply apprehensions might improve the market for new technologies and fuels, but it should be kept in mind that synthetic fuels based on coal or natural gas are more available and generally less costly. Also, the Annex 1 countries would follow sterner climate policies over the next several decades without any developing country contribution. This might entail a more rapid transition in the transportation sector if they put on a higher carbon price than other countries do.

References

[1] Façanha, C; Blumberg, K; Miller, J. *Global transportation energy and climate roadmap, The International Council on Clean Transportation*, Nov 2012.

[2] *Global Transport Scenarios 2050*, World Energy Council, 2011.

[3] Road Transport: The Cost of Renewable Solutions, *International Renewable Energy Agency (IRENA) Report*, 2013.

[4] Koushki, PA; Al-Fadhala, S; Al-Saleh, O; Aljassar, AH. Urban Air Pollution Impact of Modal Shift in School Transportation in Kuwait, *Journal of urban planning and development*, 2002, 128(2), 89-104.

[5] Fontaras, G; Zacharof, N; Ciuffo, GB. Fuel consumption and CO_2 emissions from passenger cars in Europe – Laboratory versus real-world emissions, *Progress in Energy and Combustion Science*, Vol. 60, 2017, pp. 97-131.

[6] Oliver-Hoyo, MT; Pinto, G. Using the Relationship between Vehicle Fuel Consumption and CO2 Emissions To Illustrate Chemical Principles, *Journal of Chemical Education*, Vol. 85, No. 2, 2008, pp. 218-220.

[7] Alkandari, AM; Murad, MM. Consumption comparison of different modes of operation of a hybrid vehicle at Kuwait, *International journal of advanced research in engineering*, Vol. 2(2), 2016, pp. 13-16.

[8] Ettouney, RS; Wahab, SA; Elkilani, AS. 2009. Emissions inventory, ISCST, and neural network modelling of air pollution in Kuwait, *International Journal of Environmental Studies*, 66, 2, 193-206.

[9] http://cdiac.ess-dive.lbl.gov/trends/emis/kuw.html.

[10] Al-Bassam, E; Khan, A. Air pollution and road traffic in Kuwait, *Urban Transport X*, 2004 WIT Press, pp. 741-750.

[11] Steven, H; Mulawa, C; Hunsanger, P; Nelson, EC; Ragazzi, KRA; Barrett, R. 2011, Light-Duty Motor Vehicle Exhaust Particulate Matter Measurement in the Denver, Colorado, Area, *Journal of the Air & Waste Management Association*, Vol. 49, (1999), (9), pp. 164-174.

[12] http://luc.devroye.org/kuwait.html.

[13] Bongardt, D; Schaltenberg, P. Transport in a green economy. Transport policy advisory services, *Federal Ministry for economic cooperation and development*, 2011, pp. 1-4.

[14] Hilal A. Al-Ismaily, Douglas Probert, Transport policy in Oman, *Applied Energy*, Vol. 61, Issue 2,1998, Pages 79-109.

In: Energy Policy
Editor: K. J. Sreekanth
ISBN: 978-1-53613-744-6

Chapter 2

DO THE COSTS OF RENEWABLE ENERGY PROMOTION OUTWEIGH ITS BENEFITS? AN ANALYSIS OF THE EUROPEAN CASE

Margarita Ortega Izquierdo[1], PhD and Pablo del Río[2,*], PhD
[1]Universidad de Burgos, Burgos, Spain
[2]Consejo Superior de Investigaciones Científicas (CSIC), Madrid, Spain

ABSTRACT

Major changes in behavior and technologies will be required in order to reach the ambitious decarbonisation goal of the Paris Agreement. Innovation and diffusion in low-carbon technologies will contribute to this goal. Among these technologies, renewable energy technologies have a main role to play. Indeed, it is widely acknowledged that electricity from renewable energy sources (RES-E) brings many environmental and socioeconomic benefits for society compared to conventional electricity. The comparatively lower social costs of these technologies (once their socioeconomic and environmental benefits are taken into account), their higher private costs, and the cost reductions that they experience with

* Corresponding Author Email: pablo.delrio@csic.es.

increased diffusion have been the reasons for their public promotion everywhere. However, such support has increased substantially in many countries around the world, raising the concerns of governments about the social acceptability and political feasibility of the policies. This has called into question the continuation of RES-E promotion. However, in so far as the socioeconomic and environmental benefits are high, RES-E support costs can be justified. The aim of this paper is to identify the main socioeconomic and environmental benefits of RES-E and to compare them with RES-E support costs, using a novel methodology developed by the authors in the past. The analysis is geographically circumscribed to the Member States of the European Union. It is found out that, indeed, the socioeconomic and environmental benefits in terms of CO_2 emissions reductions and fossil fuel savings are substantial in the analysed period (2010-2015). In monetary terms, the fossil fuel savings are more relevant than the CO_2 emissions savings. While the monetary calculation of the avoided CO_2 emissions has experienced an increasing trend over the period, the evolution of fossil fuel savings reached a maximum in 2013. The unitary support costs (€/MWh) increased over the period. Our results show that the total support costs have been above the socioeconomic and environmental benefits, although it should be taken into account that benefits and costs fall on different agents and, thus, they are not directly comparable.

Keywords: renewable energy electricity, Europe, support mechanisms, CO_2 emissions, fossil fuel savings, import savings

INTRODUCTION

The decarbonisation of the energy system should be a main priority of governments everywhere in order to avoid catastrophic interference with the climate system. The Paris Agreement was an important step forward in this regard. It brought all nations into a common cause to undertake ambitious efforts to combat climate change. Its central aim was to strengthen the global response to the threat of climate change by keeping a global temperature rise this century well below 2.0 degrees Celsius above pre-industrial levels and to pursue efforts to limit the temperature increase even further to 1.5 degrees Celsius [1].

To reach these ambitious goals, major changes in behavior and technologies will be required. Innovation and diffusion in low-carbon technologies will contribute to the goal of the Paris Agreement. Among these technologies, renewable energy technologies have a main role to play. Indeed, there has been a worldwide trend towards the deployment of these technologies.

It is widely acknowledged that electricity from renewable energy sources (RES-E) brings many environmental and socioeconomic benefits for society compared to conventional electricity. However, since its costs are still generally higher than those of its conventional counterpart, it could not penetrate the electricity market without public support. In addition, these technologies, as any other technology, were initially very expensive because they had not advanced along their respective learning curves. Once a technology diffuses, its costs go down due to dynamic economies of scale and learning effects. The comparatively lower social costs of these technologies (once their socioeconomic and environmental benefits are taken into account), their higher private costs and the aforementioned dynamic argument (cost reductions with increased diffusion) have been the reasons for the public promotion of RES-E everywhere. In particular, targets for renewable energy were set in the EU for 2010, 2020 and 2030 and RES-E promotion schemes have been adopted in all countries since the 1990s.

However, such support has increased substantially in many countries around the world [2], raising the concerns of governments about the social acceptability and political feasibility of the policies. These policies have usually been based on administratively-set feed-in tariffs (FITs) or premiums (FIPs), which guaranteed prices for RES-E generators, usually without a cap on the total support costs.

Concerns about the increasing support costs have called into question the continuation of RES-E promotion. However, in so far as the socioeconomic and environmental benefits are high, RES-E support costs can be justified. The aim of this paper is to identify the main socioeconomic and environmental benefits of RES-E and to compare them with RES-E support costs. The analysis is geographically circumscribed to the Member States (MS) of the European Union (EU).

Accordingly, this chapter is structured as follows. The next section provides an overview of the current situation of RES-E generation in Europe as well as information on the RES-E support instruments which are applied. The next section discusses the methodology used in this paper in order to calculate the benefits of RES-E generation and the costs of support over the analysed period (2010-2015). Then, those benefits and costs are calculated and compared. The chapter closes with some concluding remarks.

CURRENT SITUATION IN EUROPE

Renewable Electricity Generation

Net electricity production in the EU amounted to 3,066 TWh in 2015 [3]. It was concentrated in a few countries. Germany, France, United Kingdom, Spain and Italy accounted for 66% of electricity generation.

During the last decade, the EU has diversified its electricity generation mix. Several energy sources has significant shares in this mix, including coal (25.1%), nuclear (26.5%), natural gas (15.1%) and renewable energy sources (30.8%). Oil has a negligible share in electricity generation (2.6%), although it is the dominant fuel in Malta and Cyprus. Within RES, wind electricity generation has increased substantially in the last years. It has a share of 9.3%, although it is below the share of hydro (11.5%). Biomass, biogas and waste (5.5%), solar (3.3%) and geothermal (0.2%) are growing fast but have lower shares [3].

RES-E significantly increased between 2010 and 2015 in the EU, with an average annual growth rate of 15% (excluding large hydro). Hydropower, which is still the dominant RES-E, is relatively less important than it was a decade ago (Figure 2). This is due to the substantial increase in the deployment of other RES-E technologies, such as on-shore wind and biomass. In turn, this increase has been triggered by national support schemes, in particular for solar photovoltaics (PV) and wind on-shore. Other factors have also played an important role, including the constant technological improvement and intense supplier competition which have led to significant reductions in the costs of renewable energy technologies [4].

In relative terms, electricity generation from solar photovoltaics (PV) has experienced the highest growth rate (a 70-fold increase over the last decade). Wind energy generation has increased four-fold over the same period. The rest of renewable energy sources have had lower growth rates.

The contribution of renewable energy sources to the electricity mix has increased over the last decade in virtually all the Member States. Only Latvia has reduced the contribution of renewables to its mix. This has been due to a reduction in hydro generation. The increase in the contribution of RES-E in Lithuania (from 6% in 2005 to 52% in 2015), Luxembourg (from 24% in 2005 to 68% in 2015) and Denmark (from 28% in 2005 to 67% in 2015) is worth mentioning.

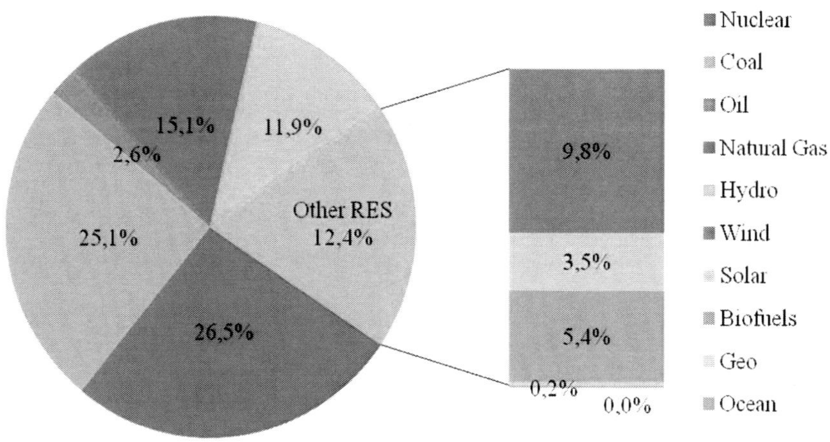

Figure 1. Net RES-E production per technology in Europe, 2015 (GWh) Source: Eurostat, 2017 [3].

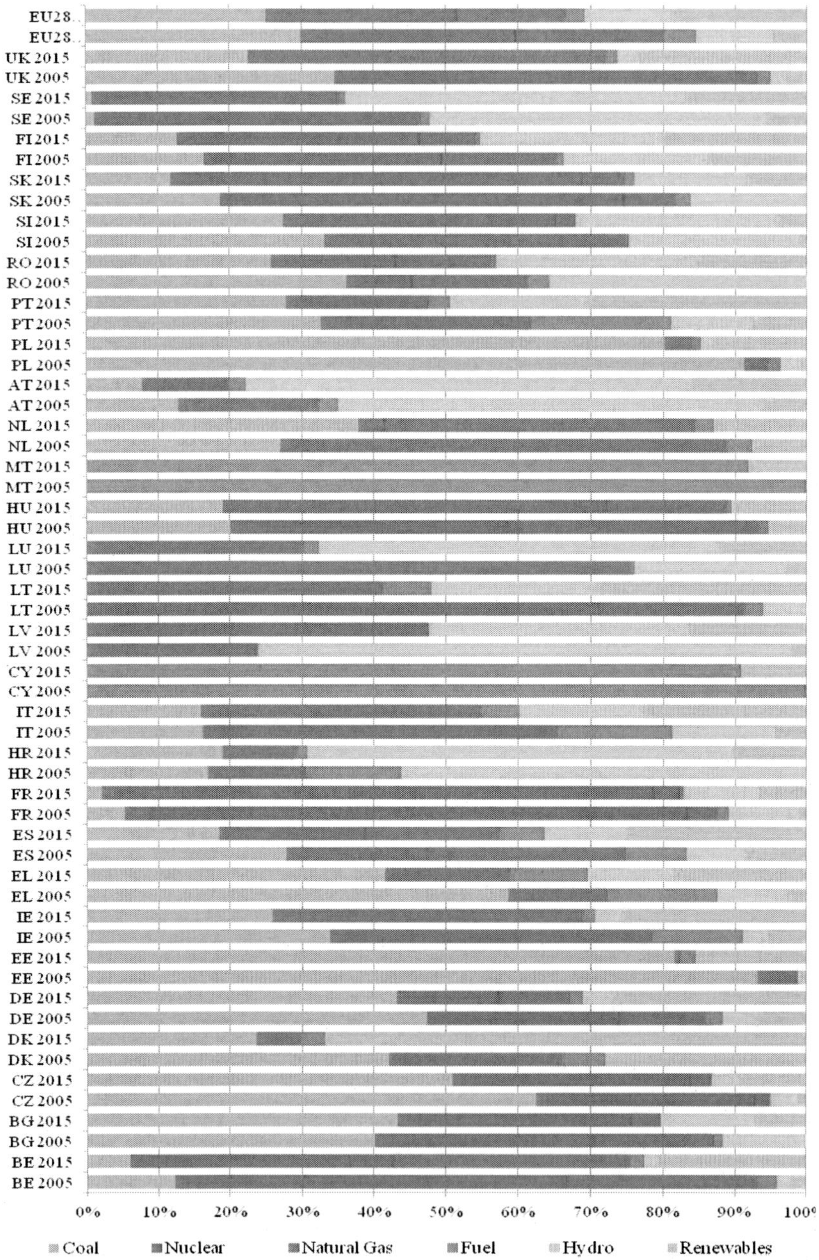

Figure 2. Evolution of the electricity mix per country. Source: Eurostat, 2017 [3].

Renewable Electricity Support Schemes

Due to the higher investment costs of RES-E, support schemes have been adopted in all MS and they have played a crucial role in RES-E deployment. The analysis of the performance of RES-E support schemes in the EU has been the focus of much research in the last decade (see, for example, [5-10]).

Basically, schemes to support RES-E can be split into three types [11]:

- *Price-base instruments.* Two options are available: feed-in tariffs (FITs), whereby a fixed price per kWh electricity is provided at a guaranteed level (irrespective of the wholesale price). Feed-in premiums (FIPs) are paid on top of the wholesale price for electricity. Premiums are calculated as the difference between the average wholesale price and a previously defined guaranteed price. In some countries, if the wholesale price rises above the guaranteed price, generators can be required to pay back the difference.
- *Quantity-base instruments.* Tradable green certificates (TGC) with quota obligations require that a percentage of electricity is generated with renewable energy sources. In tendering schemes (TN), developers compete for supply contracts to build RES-E capacity. Tenders are often coupled with other types of support e.g., FITs, FIPs or green certificates.
- *Other instruments* include tax exemptions (TE), investment incentives (II) and financial incentives (FI).

In 2014, Spain implemented an innovative system based on an investment and operation return for existing plants, calculated in accordance to a reasonable Internal Rate of Return (IOR)

As Table 1 shows, FIT is the dominant support scheme in most Member States, although its importance has diminished in the last years. FIPs have experienced the opposite trend, and they have been increasingly adopted, especially for the large installations. This has mostly been due to the European Commission recommendation to use more "market-based" cost-efficient instruments. Indeed, competitive auctions have to be implemented in order to provide support to all new installations from 2017 onwards [17]. Some countries have already implemented auctions as their primary support scheme, including Germany, Spain, U.K., France, Italy, the Netherlands, Denmark and Poland. Those "primary" instruments are sometimes combined with "secondary" ones, including investment subsidies, soft loans, tax exemptions or net metering [18].

A reduction in public support has been experienced in the last years. This has been the result of the fast technological development of renewable energy technologies (leading to cost reductions), the achievement of interim targets and the economic/financial crisis. Germany changed to a FIP support scheme in 2014, Greece drastically reduced its remuneration levels under FITs (by around 25%), the Czech Republic cancelled public support to new solar PV and biogas plants in 2014, Spain abolished its previous support scheme and, since 2013, a new support scheme applies to existing plants, whereby they receive investment-based support (instead of generation-based support, as it was the case before).

Table 1. Instruments for RES-E deployment in the EU (2015). Source: CEER, RES-Legal and own elaboration [12-16]

	Primary instrument support					Secondary instrument support				
MS	*Hydro*	*Wind*	*Bio*	*Solar*	*Geo*	*Hydro*	*Wind*	*Bio*	*Solar*	*Geo*
AT		FIT	FIT	FIT(1)	FIT				II(2)	
BE	TGC	TGC	TGC	TGC	TGC	NM; II	NM; II	NM; II	NM;II	NM; II
BG			FIT	FIT						
CY				FIT					II	
CZ	FIT(4); FIP(4)	FIT(4); FIP(4)	FIT(3)(4); FIP(3)(4),	FIT(3); FIP(3)	FIT(4); FIP(4)	II				
DE	FIT; FIP(5)	FIT; FIP(5)	FIT; FIP(5)	FIT; FIP(5)	FIT; FIP(5)	FI	FI	FI; II	FI; TN	FI
DK	FIP	FIP; TN(6)	FIP	FIP	FIP		FI(7);			
EE	FIP	FIP	FIP	FIP	FIP		II	II		
EL	FIT	FIT	FIT	FIT	FIT	NM	NM	NM	NM	
ES	IOR	IOR	IOR	IOR	IOR					

	Primary instrument support					Secondary instrument support				
MS	*Hydro*	*Wind*	*Bio*	*Solar*	*Geo*	*Hydro*	*Wind*	*Bio*	*Solar*	*Geo*
FI		FIP	FIP			II	II	II	II	II
FR	FIT	FIT(8); TN(9)	FIT	FIT(10); TN(11)	FIT	TE	TE	TE	TE	
HR	FIT	FIT	FIT	FIT		FI	FI	FI	FI	FI
HU	FIT	FIT	FIT	FIT	FIT	NM	NM	NM	NM	NM
IE	FIT	FIT	FIT							
IT	TGC(12); FIT(13)	TGC(12); FIT(13)	TGC(12); FIT(13)	FIT(14); FIP(15)(16)	TGC(12); FIT(13)	TE; TN	TE; TN	TE; TN	TE	TE; TN
LT	FIP; TN(17)	FIP; TN(17)	FIP; TN(17)	FIP; TN(17)		II, FI	II, FI	II, FI	II, FI; NM	II, FI
LU	FIT(18)	FIT	FIT(19)	FIT(20)		II	II	II	II; TE	II
LV	FIT	FIT	FIT	FIT	FIT	NM	NM	NM	NM	NM
MT				FIT(21)					II	
NL	FIP	FIP	FIP	FIP; TN(22);	FIP	FI; TE	FI; TE		FI; TE; NM	FI; TE
PL	TGC	TGC	TGC	TGC	TGC					
PT	FIT	FIT	FIT	FIT	FIT					
RO	TGC	TGC	TGC	TGC	TGC	II	II	II	II	II
SE	TGC	TGC	TGC	TGC	TGC		TE	TE	II; TE	TE
SI	FIT	FIT	FIT	FIT	FIT	II; FI	II; FI	II; FI	II; FI	II; FI
SK	FIT	FIT	FIT	FIT(23)	FIT	II, TE	II(24), TE	II, TE	II(24), TE	II, TE
UK	FIP/ TN(25) TGC; FIT(26)	FIP/TN(25) TGC; FIT(26)	FIP/TN(25) TGC; FIT(26)	FIP/TN(25) TGC; FIT(26)	FIP/TN(25 TGC; FIT(26))	TE	TE	TE	TE	TE

(1): Solar PV – >5kW; (2): Solar PV – ≤5kW; (3): New PV installations and biogas plants are only being supported if put into operation before 31 December 2013; (4): Wind, hydro, geothermal and biomass plants that were put into operation before 31 December 2015 and their building permit was issued before 2 October 2013 are eligible for support; (5): Mandatory FIP scheme for all new installations, since 08/2014; (6): Wind offshore; (7): Wind onshore; (8): Wind onshore; (9): Wind offshore; (10): Solar PV – Included P<100kW; (11): Solar PV – Included P>100kW; (12) New plants up to 31 December 2012. From 2016 TGCs have been substituted by FIPs on electricity produced; (13) New plants up to 1 MW before 29 June 2016 (500 kW from 30 June 2016): FITs on electricity injected into the grid; (14) New plants from 27 August 2012 to 6 July 2013 - up to 1 MW; (15) New plants up 26 August 2012; (16) New plants from 27 August 2012 to 6 July 2013 (> 1 MW); (17): >10kW; (18): <6MW; (19): Biogas <2,500kW and biomass <10MW; (20): <30kW; (21): <1MW; (22): Wind offshore; (23): PV – <30kW; (24): <10kW; (25) Contracts for difference scheme >5MW; (26) <5MW.

FIT: Feed-in Tariff; *FIP:* Feed-in Premium; *TGC:* Tradable Green Certificates; *TN:* Tenders; *II:* Invest Incentive; *FI:* Financial Incentive; *TE:* Tax Exemption; *NM:* Net Metering.

Member States abbreviations: AT: Austria; *BE:* Belgium; *BG:* Bulgaria; *CZ:* Czech Republic; *CY:* Cyprus; *DE:* Germany; *DK:* Denmark; *EE:* Estonia; *EL:* Greece; *ES:* Spain; *FI:* Finland; *FR:* France; *HR:* Croatia; *HU:* Hungary; *IE:* Ireland; *IT:* Italy; *LT:* Lithuania; *LU:* Luxembourg; *LV:* Latvia; *MT:* Malta; *NL:* Netherlands; *PL:* Poland; *PT:* Portugal; *RO* Romania; *SE:* Sweden; *SI:* Slovenia; *SK:* Slovakia; *UK:* United Kingdom.

PV, wind onshore, hydropower and bioenergy are the most widely supported RES. Support for other renewable energies such as wind offshore, geothermal and Concentrated Solar Power (CSP) is less widespread.

METHODOLOGY[1]

The deployment of RES-E brings several benefits to society. RES-E deployment leads to two main benefits: CO_2 emissions reductions and lower fossil-fuel consumption levels. These can be quantified in monetary terms. There are other benefits related to RES-E deployment, including other environmental benefits (reductions of local pollutants, such as SO_2), the merit-order effect, which refers to the reduction of the wholesale electricity price due to a greater penetration of RES-E and industry and job creation. Given the lack of reliable EU-wide data, they have not been considered in this chapter.

With respect to the cost side, only policy support costs have been considered. System generation costs have not been taken into account. These would be needed in order to carry out a complete cost-benefit analysis. The costs for the electricity system related to the variable character and the limited predictability of the generation of electricity from intermittent RES-E (usually called cycling costs and balancing costs) have not been included because their values are negligible compared to other impacts. This is so at least with low RES-E penetration levels, as it has been argued by [20] and [21].

The two aforementioned benefits associated to RES-E deployment have been quantified using a methodology developed by the United Nations Framework Convention on Climate Change (UNFCCC) [22]. This methodology was not originally conceived for this purpose and it is commonly used in the calculation of baselines in clean development

[1] This section heavily draws from [19].

mechanism (CDM) projects. The methodology was adapted by [19], who applied it to the Spanish case.

According to this methodology, CO_2 emission factors can be obtained using three different alternatives:

1) *Operating margin factor (OM).* It considers the existing power plants. Their current electricity generation would be affected by the proposed RES-E project. The so-called Average OM alternative will be used in this paper in order to calculate the operating margin. The Average OM emission factor is calculated as the generation-weighted average CO_2 emissions per unit of net electricity generation of all generating power plants serving the system, including low-cost/must-run power plants[2].

According to [22], electricity imports from other countries are treated as an additional electricity generation technology.

The calculation of the Average OM emission factor is based on the following equation:

$$EF_{\text{Average OM},y} = \Sigma_m\, EG_{m,y} \cdot EF_{EL,m,y})/\Sigma EG_{m,y} \qquad \text{(Eq. 1)}$$

$EF_{\text{Average OM},y}$ refers to the operative margin factor in the year *y* (tCO_2/MWh); $EG_{m,y}$ represents the net electricity generated by the technology *m* in the year *y* (MWh), $EFE_{Lm,y}$ refers to the CO_2 emission factor of technology *m* in the year *y*; *m* are the technologies serving the system (according to the Average OM definition, these are all the power generation technologies in the system) and *y* is the year of reference.

2) *Build margin factor (BM).* This refers to prospective power plants whose construction would have been delayed or cancelled by the increase in RES-E generation. An intermediate scenario has been considered due to the lack of data at European data on the installed capacity per fuel type. In this scenario, a 50% substitution of coal and natural gas is assumed. As coal and natural gas thermal plants are not located in neither Cyprus nor Malta, a single scenario in which RES-E replaces fuel-oil is considered for

[2] Low-cost/must-run resources are power plants with low marginal generation costs or dispatched independently of the daily or seasonal grid load. Both nuclear and renewable energies (including hydro) belong to this category.

these two countries. The BM emission factor will be used as the emission factor of the replaced technology or technologies.

3) *Combined margin factor (CM).* This is a weighted average of the two preceding factors. The OM and the BM factors have equal weights:

$$EF_{CM,y} = 0{,}5 \cdot EF_{Average\ OM,\ y} + 0{,}5 \cdot EF_{BM,\ y} \qquad \text{(Eq. 2)}$$

where $EF_{CM,y}$ refers to the CM factor in the year *y*, $EF_{Average\ OM,y}$ is the OM factor (using the average OM alternative) in the year *y* and $EF_{BM,\ y}$ refers to the BM factor in the year *y*.

Thus, the impact of renewable energy plants on the existing power plants (OM) and the avoided construction of new facilities (BM) is considered in this analysis.

Benefits from the Avoided CO_2 Emissions

The calculation of the CO_2 emission reductions related to RES-E deployment is based on the application of the original UNFCCC methodology. [23] provides the emissions from fossil fuel combustion. The oxidation factors for the three fuels considered, which are those recommended by [24], are the following: coal (99%), fuel-oil (99.5%) and natural gas (99.5%). The emissions factors are: coal (97.3 tCO_2/TJ), fuel-oil (77 tCO_2/TJ) and natural gas (55.8 tCO_2/TJ).

In order to translate CO_2 emissions reductions into monetary terms, the concept of the Social Cost of Carbon (SCC) has been used. The SCC reflects the external costs of greenhouse gas emissions. It measures the full global cost today of an incremental unit of carbon emitted now, summing the full global cost of the damage it imposes on the atmosphere. It measures the scale of the externality which needs to be incorporated into decisions on policy and investment options in government [25]. The SCC has very different values depending on the study being considered, as shown in the review made by [26]. The median of the values of the SCC in [26] has been used (32 €/tCO_2) in this chapter. This value is higher than the price of CO_2 allowances under the EU Emission Trading Scheme (EU ETS) in the reference period being considered (a maximum of 14€/tCO_2 in 2010 and a minimum of 4.5€/tCO_2 in 2013). However, it could be regarded as a reasonable scenario for the SCC, being the central value of the range analysed by [26]. Obviously, the monetary calculations of the avoided emissions would vary if the mode (14 €/tCO_2) or the average (49 €/tCO_2) were applied.

Benefits from Fossil Fuels Savings

Final energy has been transformed into primary energy. The result has been multiplied by the price of energy in order to calculate the fossil fuel savings as a result of RES-E deployment. The original UNFCCC methodology is adapted in order to calculate the conversion from final to primary energy and the price of energy. The emission factors of the technologies are replaced by the efficiency values and prices of these technologies, respectively.

The efficiency values considered for the transformation of primary energy into final energy have been obtained from [27,28]. It has been assumed that the efficiency of fossil fuel plants does not change when they are substituted by RES-E. However, a very small increase in those efficiencies could be expected in reality [21].

The following equation is used to transform final energy into primary energy for both the OM and BM alternatives:

$$1 \text{ MWh (Primary energy)} = 1/\eta \text{ MWh (Final energy)} \qquad \text{(Eq. 3)}$$

The emission factors in the OM and BM alternatives are replaced by the efficiency values of the considered fuels in order to calculate the conversion factors of final energy into primary energy in the CM alternative.

Table 2. Combined margin factors used: conversion factor (final energy to primary energy), emission factors and prices. Years 2010 and 2015. Source: Own elaboration

MS	$MWh_{final}/MWh_{primary}$		tCO_2/MWh_{final}		€/$MWh_{primary}$	
	2010	*2015*	*2010*	*2015*	*2010*	*2015*
AT	1.75	1.69	0.37	0.34	16.3	14.2
BE	2.23	2.05	0.38	0.37	13.2	13.4
BG	2.21	2.14	0.49	0.47	12.9	11.2
CY	2.26	2.21	0.62	0.60	36.4	28.2
CZ	2.25	2.17	0.52	0.48	13.5	11.9
DE	2.16	2.08	0.49	0.49	14.3	12.1
DK	1.98	1.80	0.48	0.38	17.3	12.9
EE	2.17	1.96	0.64	0.51	16.4	14.9
EL	1.99	1.90	0.54	0.49	18.5	16.2
ES	2.01	2.03	0.41	0.42	14.2	12.4
FI	2.14	2.06	0.41	0.35	13.7	10.7
FR	2.36	2.37	0.33	0.32	10.0	8.5
HR	1.67	1.65	0.36	0.35	21.5	19.4
HU	2.16	2.04	0.40	0.36	14.5	13.5
IE	1.99	1.93	0.50	0.48	18.1	15.5
IT	1.94	1.92	0.46	0.43	18.4	15.4
LT	1.70	1.68	0.35	0.34	25.9	22.0
LU	1.68	1.61	0.35	0.32	24.0	20.5
LV	1.67	1.73	0.35	0.34	20.6	17.6
MT	2.26	1.95	0.63	0.47	36.6	31.7
NL	2.04	1.99	0.49	0.49	18.0	15.6
PL	2.18	2.13	0.65	0.61	16.2	13.3
PT	1.84	1.93	0.41	0.46	16.0	14.4
RO	2.02	1.95	0.45	0.42	13.2	11.8
SE	2.03	1.97	0.31	0.31	10.8	8.4
SI	1.93	1.92	0.38	0.37	16.0	13.5
SK	2.09	2.03	0.36	0.34	13.2	12.2
UK	2.15	2.11	0.51	0.44	15.7	12.6
EU28	2.12	2.07	0.44	0.42	14.2	12.0

The prices of fossil fuels used in electricity generation have been obtained from [29]. Electricity imports have been treated as if they were an independent technology generator and data on the wholesale electricity prices from each MS for the period analyzed are provided by [30]. For those countries without available data and for non-EU MS, the average of the electricity price of the rest of countries (i.e., 43.5 €/MWh in 2015) has been used. The following standard conversion factors have been applied: Coal: 1MWh = 0.21 t; Oil: 1 MWh = 0.61 barrels; Natural gas: 1 MWh = 3.44 · 106 Btu. [31]

According to data provided by the International Monetary Fund [32], the average price of uranium in the period 2010-2015 was 43.3 €/pound. Assuming the unitary fuel consumption in [33], the price of fuel for nuclear electricity generation was 0.23 €/MWh. Compared to the price of other fossil fuels, this price is negligible during the analyzed period and, thus, it was not included in the study.

Finally, it has been assumed that biomass, biogas and wastes are indigenous energy sources and, thus, that there are not any imports.

Table 2 shows the conversion factor from final to primary energy, the CO_2 emission factor (tCO_2/GWh_{final}) and the price ($€/MWh_{primary}$) that is obtained when this methodology is applied.

Support Cost Calculation

Support costs, defined as the annual cost of incentives paid to RES-E generation as the result of national support schemes, have been obtained from [12-16].

RESULTS AND DISCUSSION

CO_2 Emissions Avoided

The methodology described above has been applied to calculate the CO_2 emissions avoided with RES-E generation.

Figure 3 shows the economic savings due to CO_2 emissions reductions. RES-E prevented the emission of 2043 million tons of CO_2, leading to savings of 65384M€ over the 2010-2015 period.

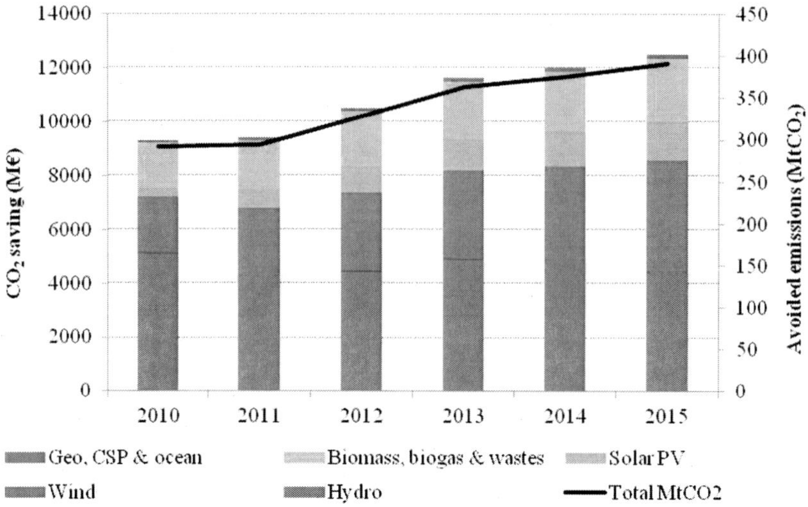

Figure 3. Evolution of avoided CO_2 emissions and economic savings in the EU between 2010 and 2015. Source: Own elaboration.

Germany, Italy and Spain were the main contributors to the aforementioned economic savings. Regarding the contribution of the different renewable energy technologies to CO_2 savings, hydro has experienced a slight reduction. In contrast, the contributions of wind, biofuels and PV have substantially increased over the period. In 2015, the shares of each technology to emissions reductions were as follows: 35% for hydro, 33% for wind, 19% for biofuels, 12% for solar PV and 1% for the rest of technologies.

Fossil Fuel Savings

RES-E generation has led to substantial energy savings in the EU. 10000 TWh of fossil-fuel generation were replaced with RES-E generation in the 2010-2015 period. Hydro and wind contributed most to those savings. However, the savings due to hydro have remained more or less constant in the period, whereas those related to other renewable energy technologies have increased. Wind and solar have been particularly relevant in this regard, since their contributions to those savings have substantially increased during the period, as shown in Figure 4.

This energy saving entailed an economic saving of 150000M€ for the EU. Germany, Italy, Spain and France had the largest economic savings.

The evolution of energy savings due to the penetration of RES-E has increased annually during the period being analyzed (Figure 4). However, the economic savings reached a maximum in 2013 and experienced a reduction afterwards (in 2014 and 2015). The unitary price of primary energy is linked to the price of fossil fuels and the electricity mix in each country. The low prices of fossil fuels in 2014 and 2015 are behind the reduction of the economic savings in those years.

The economic savings in fossil fuels are greatest in countries with a stronger energy dependence since they reduce their fossil-fuel imports. 90000 M€ of fossil-fuel imports were saved in the period 2010-2015 in the EU. Italy, Germany, Spain, France and Sweden benefited most from reductions in fossil-fuel imports. Table 3 shows the economic savings due to the reduction of imports as a result of RES-E generation in the EU.

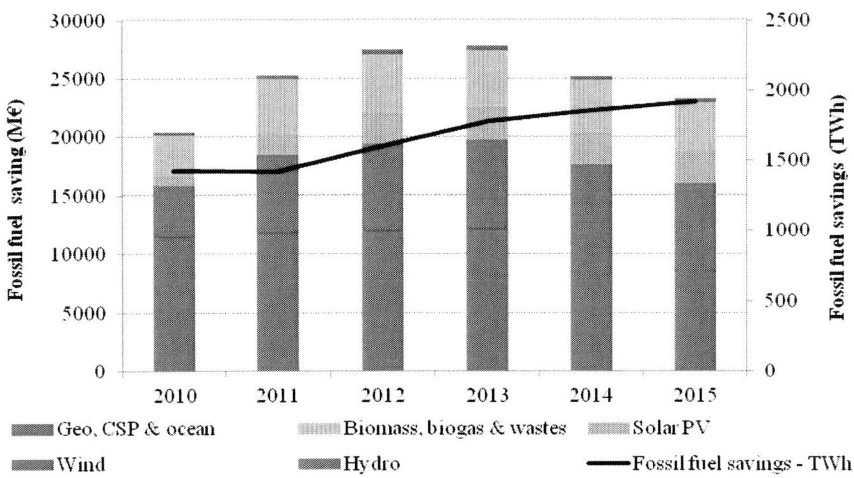

Figure 4. Evolution of economic savings due to the reduction in fossil-fuel consumption in the EU between 2010 and 2015. Source: Own elaboration.

Table 3. Economic savings due to the reduction of imports in the EU (2010-2015). Source: Own elaboration.

Imports savings (M€)	2010	2011	2012	2013	2014	2015
Hydro	6604	7176	7628	7640	6833	5434
Wind	2460	3873	4378	4553	4348	4534
Solar PV	379	981	1434	1563	1588	1539
Biofuels	1967	2732	3003	2906	2731	2524
Geo, CSP & ocean	103	165	198	198	193	176
TOTAL	11513	14926	16641	16859	15693	14206

Public Support Cost

As explained in Section Two, virtually all the Member States of the European Union have implemented some type of RES-E support scheme. Data on public support are available for the 2010-2015 period for all countries except Bulgaria and Slovakia. Data for the Netherlands are not available for 2014 and 2015 [15].

Table 4 shows the evolution of the unitary cost of public support in the EU.

In general, the unitary public support has increased slightly during the period for all renewable energy technologies, with the exception of solar PV. This technology receives the highest unitary public support (annual average: 331.87€/MWh), but substantial reductions of this support have been experienced in the last years. The reason behind this reduction is the achievement of the targets for this technology as set in the National Renewable Energy Action Plans (NREAPs) of the Member States, the reductions in the costs of this technology and the high total support costs for this technology in some countries. The reductions during the period have been substantial in Belgium (57%), Slovenia (43%), Sweden (34%), Germany, Italy and France (29%). In several Member States (e.g., the Czech Republic), the new solar PV installations do not receive any public support.

Table 4. Evolution of unitary public support for RES-E in the EU (2010-2015) Source: CEER [12-15]

Annual average support (€/MWh)	2010	2011	2012	2013	2014	2015
Hydro	43.34	45.74	48.30	55.71	67.52	64.94
Wind	48.98	48.15	52.15	55.92	55.93	60.44
Solar PV	396.02	366.96	347.00	318.59	291.28	271.40
Biofuels	83.54	91.03	88.89	105.75	104.41	107.21
Geo, CSP & ocean	193.48	242.41	210.94	193.00	204.88	206.07
TOTAL	87.16	102.43	107.70	111.93	112.42	110.36

Wind and hydro usually receive the lowest support levels (an average of 53.9 and 54.2€/MWh, respectively) in the EU Member States. Bioenergy is in an intermediate position (annual average of 107.21€/MWh).

Notwithstanding, even for the same technology, support levels widely differ per country, although these differences are not always related to renewable resource potentials, a result in line with [34].

PUBLIC SUPPORT COSTS VS BENEFITS OF RES DEPLOYMENT

Figure 5 shows the evolution of savings related to RES-E generation as well as the trend of the total support received by the renewable energy technologies in the EU. The results show that the costs of public support are higher than the benefits. Recall that benefits and costs fall on different agents and, thus, they are not directly comparable. Benefits are enjoyed by citizens, while costs usually fall on electricity consumers.

Fossil fuel savings are much higher than the savings related to CO_2 emissions reductions (both savings are calculated in monetary terms), a result in line with [20].

However, behind this broader picture, significant country and technology differences exist. Benefits are above the policy costs for hydro, but they are below those costs for bioenergy, PV and other RES-E (geothermal, solar thermoelectric and ocean). The benefits and costs are similar in the case of wind.

Albeit PV is already a mature technology and its costs have significantly been reduced in the analyzed period, it was still an expensive one, at least during such period. Consequently, support levels for PV have been considerably higher than the associated benefits in all countries. In contrast, the benefits of hydro outweigh the costs of support in all countries.

During the period, the costs of public support for wind energy reached 65472 M€ and the benefits were 59920M€. The benefits of wind energy exceeded the costs of support in 16 Member States. The cases of Spain, Denmark and Ireland are worth mentioning in this regard.

The support costs of biofuels exceed the benefits in the period for the EU. However, the benefits are higher than the costs in 15 Member States, especially in Finland, Poland and Sweden.

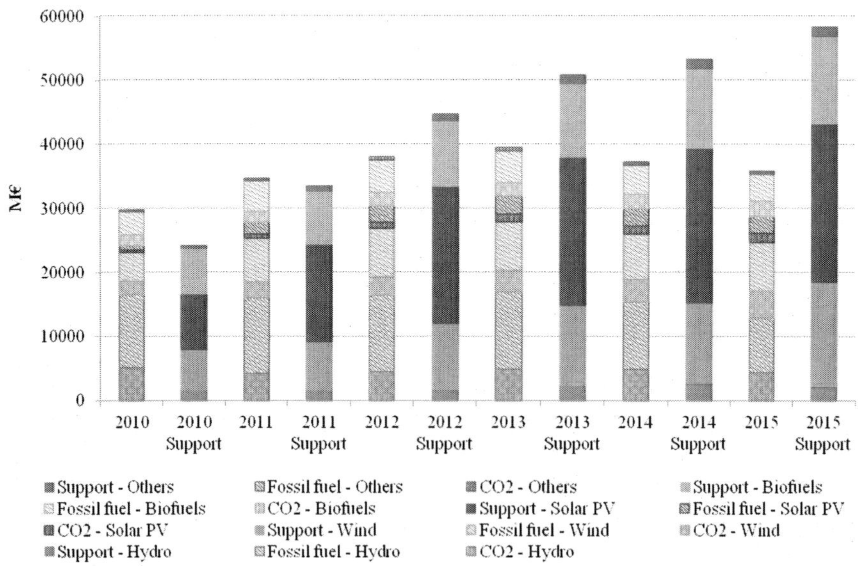

Figure 5. Evolution of the costs and benefits of renewable energy generation in the EU. Source: Own elaboration.

Conclusion

It is widely acknowledged that electricity from RES-E brings many environmental and socioeconomic benefits for society compared to conventional electricity and, thus, support is justified. However, such support has increased substantially in many countries around the world, including the EU Member States, raising the concerns of governments about the social acceptability and political feasibility of the policies. This paper has identified the main socioeconomic and environmental benefits of RES-E using a novel methodology developed by the authors in the past. Those benefits have been compared with RES-E support costs.

It is found out that, indeed, the socioeconomic and environmental benefits in terms of CO_2 emissions reductions and fossil fuel savings are substantial in the analysed period (2010-2015). However, there are some differences between the two types of benefits. First, the fossil fuel savings are more relevant than the CO_2 emissions savings in monetary terms. Second, the evolution of those benefits differs across the analysed period. While the monetary calculation of the avoided CO_2 emissions has experienced an increasing trend over the period, the evolution of fossil fuel savings reached a maximum in 2013. Third, the unitary support costs (€/MWh) increased over the period. Fourth, our results show that the total support costs have been above the socioeconomic and environmental benefits. However, it should be taken into account that benefits and costs fall on different agents and, thus, they are not directly comparable.

It should be taken into account that RES-E deployment brings several benefits in addition to CO_2 emissions reductions and fossil fuel savings which have not been included in this analysis, including industry and job creation, export opportunities and rural and regional development. Further research could aim to calculate those benefits in monetary terms and add them to our calculations.

REFERENCES

[1] United Nations (2015). *Paris Agreement.* Paris, France. Available at: http://unfccc.int/files/essential_background/convention/application/pdf/english_paris_agreement.pdf.

[2] REN21 (2017). *Renewables Global Status Report 2017. Paris, France.* Available at: http://www.ren21.net/wp-content/uploads/2017/06/17-8399_GSR_2017_Full_Report_0621_Opt.pdf.

[3] EUROSTAT (2017). *Eurostat Statistical Database.* Available at: http://appsso.eurostat.ec.europa.eu/nui/show.do?dataset=nrg_105a&lang=en. Accessed 15 sep 2017.

[4] IRENA (2016). *The power to change: solar and wind cost reduction potential to 2025.* International Energy Agency. Abu Dhabi, United Emirates.

[5] Nicolini, M. and Tavoni, M. (2017). "Are renewable energy subsidies effective? Evidence from Europe". *Renewable and Sustainable Energy Reviews* 74: 412-23.

[6] Haas, R., Panzer, C., Resch, G., Ragwitz, M., Reece, G., Held, A. (2011) "A historical review of promotion strategies for electricity from renewable energy sources in EU countries". *Renewable and Sustainable Energy Reviews*, 15: 1000-34.

[7] Fagiani, R., Barquin, J., Hakvoort, R. (2013). "Risk-based assessment of the cost-efficiency and the effectivity of renewable energy support scheme: Certificate markets versus feed-in tariffs". *Energy Policy*; 55: 648-61.

[8] Jenner, S., Groba, F., Indvik, J. (2011). "Assessing the strength and effectiveness of renewable electricity feed-in tariffs in European Union countries". *Energy Policy*; 52: 385-401.

[9] Verbruggen, A., Lauber, V. (2012) "Assessing the performance of renewable electricity support instruments". *Energy Policy*; 45: 635-44.

[10] Winkler, J., Magosch, M., Ragwitz, M. (2018). "Effectiveness and efficiency of auctions for supporting renewable electricity – What can we learn from recent experiences?". *Renewable energy*; 119: 473-89.

[11] Del Río P. (2012) "Analysing the interactions between renewable energy promotion and energy efficiency support schemes: The impact of different instruments and design elements". *Energy Policy*; 38 (9): 4978-89.

[12] CEER (2011). *CEER Report on Renewable Energy Support in Europe.* Council of European Energy Regulators. Brussels, Belgium. Available at: www.ceer.eu

[13] CEER (2013). *Status Review of Renewable and Energy Efficiency Support Schemes in Europe.* Council of European Energy Regulators. Brussels, Belgium. Available at: www.ceer.eu

[14] CEER (2015). *Status Review of Renewable and Energy Efficiency Support Schemes in Europe in 2012 and 2013.* Council of European Energy Regulators. Brussels, Belgium. Available at: www.ceer.eu

[15] CEER (2017). *Status Review of Renewable Support Schemes in Europe. Council of European Energy Regulators.* Brussels, Belgium. Available at: www.ceer.eu

[16] European Commission (2017). R*ES Legal Europe.* Available at: www.res-legal.eu.

[17] European Commission (2014). *Guidelines on State aid for environmental protection and energy* 2014-2020, C(2014) 2322.

[18] Mir-Artigues P., del Río P. (2014) "Combining tariffs, investment subsidies and soft loans in a renewable electricity deployment policy". *Energy Policy*; 69: 430-42.

[19] Ortega M., del Río P., Montero E.A. (2013) "Assessing the benefits and costs of renewable electricity. The Spanish case". *Renewable and Sustainable Energy Reviews*; 27: 294-304.

[20] Marcantonini C, Ellerman A.D. (2015). "The Implicit Carbon Price of Renewable Energy Incentives in Germany". *The Energy Journal*; 36 (4): 205-30.

[21] Gutiérrez-Martín F., Da Silva-Álvarez R.A., Montoro-Pintado P. (2013) "Effects of wind intermittency on reduction of CO_2 emissions: The case of the Spanish power system". *Energy*; 61: 108-17.

[22] UNFCCC (2013). Tool to calculate the emission factor for an electricity system. Version 04.0. Oct, 2013. *United Nations Framework Convention on Climate Change.* Bonn, Germany.

[23] European Commission (2004). *Commission Decision 2004/156/EC of 29 January 2004 establishing guidelines for the monitoring and reporting of greenhouse gas emissions pursuant to Directive 2003/87/EC of the European Parliament and of the Council* (notified under document number C(2004) 130).

[24] IPCC (2006). *Guidelines for National Greenhouse Gas Inventories. Prepared by the National Greenhouse Gas Inventories Programme,* Eggleston H.S., Buendia L., Miwa K., Ngara T. and Tanabe K. (eds). IGES, Japan.

[25] Price R., Thornton S., Nelson S. (2007) The Social Cost of Carbon and the Shadow Price of Carbon. *What they are and how to use them in economic appraisal in the U.K. Department for Environment, Food and Rural Affairs.* London, United Kingdom. Available at: https://www.gov.uk/government/uploads/system/uploads/attachment_data/file/243825/background.pdf.

[26] Tol R.S.C. (2012). "A cost–benefit analysis of the EU 20/20/2020 package". *Energy Policy*; 49: 288-295.

[27] European Commission (2011). *Commission Implementing Decision 2011/877/EU of 19 December 2011 establishing harmonised efficiency reference values for separate production of electricity and heat in application of Directive 2004/8/EC of the European Parliament and of the Council and repealing Commission Decision 2007/74/EC* (notified under document C(2011) 9523).

[28] European Commission (2015). Commission Delegated Regulation (EU) 2015/2402 of 12 October 2015 in application of Directive 2012/27/EU of the European Parliament and of the Council and repealing Commission Implementing Decision 2011/877/EU

[29] BP (2017). *BP Statistical Review of World Energy 2017.* Available at: https://www.bp.com/en/global/corporate/energy-economics/statistical-review-of-world-energy.html. Accessed September, 9, 2017.

[30] European Commission (2017). *Electricity market reports.* Available at: https://ec.europa.eu/energy/en/data-analysis/market-analysis.

[31] IEA/OECD/EUROSTAT (2004). *Energy Statistic Manual.* Paris, France. Available at: https://www.iea.org/publications/freepublications/publication/energy-statistics-manual.html.

[32] IMF (2017). I*nternational Monetary Fund.* Available at: http://www.imf.org/external/np/res/commod/index.aspx. Accessed September 17, 2017.

[33] MINETAD (2017). *Annual Electricity Statistics. Ministry of Energy, Tourism and Digital Agenda.* Madrid, Spain. Available at: http://www.minetad.gob.es/energia/balances/Publicaciones/ElectricasAnuales/Paginas/ElectricasAnuales.aspx. Accessed September 23, 2017.

[34] del Río P., Tarancón M.A., Peñasco C. (2014) "The determinants of support levels for wind energy in the European Union. An econometric study". *Mitigation and Adaptation Strategies for Global Change*; 19(4): 391-410.

In: Energy Policy
Editor: K. J. Sreekanth
ISBN: 978-1-53613-744-6

Chapter 3

A REVIEW ON RESIDENTIAL ELECTRICITY PRICE POLICIES IN CHINA

Jiang Yu[1,2] and Zheng Fang[3,*]
[1]Center for Economic Development Research,
Wuhan University, China
[2]Center for Population, Resource and Environment Research,
Wuhan University, China
[3]School of Business, Singapore University of Social Sciences,
Singapore

ABSTRACT

This chapter reviews the history of residential electricity pricing in China in five stages. The first stage is a period during 1987-1997 when there was a power supply shortage, which led to an initial discussion on the feasibility of tiered electricity pricing. The second stage is from 1998 to 2003 when a surplus of electricity supply emerged and electricity pricing is used to promote the electricity consumption. Subsequently we focus discussion on the specifics and policy effects of the tiered electricity pricing implemented in three provinces Zhejiang, Fujian and

* Corresponding Author Email: fangzheng@suss.edu.sg.

Sichuan during 2004-2006. The fourth stage is a period during 2008-2010 when the expansion of the tiered electricity pricing policy is somewhat stagnated. Finally, the tiered electricity pricing is implemented nationwide since 2012, and the characteristics of the pricing policies across provinces and some policy effects are discussed in detail. The tiered pricing policies are still undergoing some adjustments and changes up till now.

1. Introduction

For a long time, in order to reflect the superiority of the socialist system, the Chinese government adopted a unified low-cost electricity policy for residential electricity. The policy is characterized by: First, residential electricity price has been lower than industrial electricity price since 1990s. It is about 90% of industrial electricity price in China, while in many other countries, residential electricity price is typically about 1.5-2 times industrial electricity price. In fact, it is using industrial electricity subsidizing residential consumption. Since 2003, China's coal and other primary energy prices continued to rise, electricity price also rose. But price adjustment for the residential electricity was much lower and less frequent than electricity price in other industries. Residential electricity prices have been stayed at a low level. Second, the residential electricity adopted a unified price. According to the statistics of State Grid Corporation of China, about 5% of high-income households consume about 24% of total residential electricity consumption and 10% of high-income households consume 33%. The result of the combination of low tariff policy and unified price is that: In China, the more electricity the household consumes, the more subsidy the household will enjoy, and vice versa. The unified low electricity price policy is not only against the equality principle, it cannot reflect the intrinsic value of the resource and promote energy conservation and environmental protection.

However, with the rapid development of China's economy, shortage of energy supply and increasing pressures of energy-saving emission reduction and environmental protection are gradually highlighted. Power shortage caused by the "power cuts" phenomenon is almost the norm in

China's economic development. To promote the construction of resource-saving and environment-friendly society, and gradually establish a fair electricity price mechanism so as to guide residents to rationally use electricity and save electricity, on the basis of summing up successful experience from domestic and foreign tiered electricity price reform, the Chinese government has passed the residential tiered electricity pricing proposed in the government work report at the Third Session of the Eleventh National People's Congress. Since 1 July 2012, the government implemented the tiered residential electricity pricing nationwide across 29 provinces (except Tibet and Xinjiang), which called an end to an era of unified low pricing for a decade.

Recalling the history of Chinese residential tiered electricity price reform, it is closely related to the power supply and demand gap. From the beginning of the discussion on the tiered price during the first large-scale power shortage in 1987, to three pilot projects in Zhejiang, Fujian and Sichuan during 2004-2006 caused by the second large-scale power shortage of 2004, to the third large-scale power shortage in 2011 which led to the final implementation of the tiered price in 2012, and the fine-tune of the tiered price policy during 2013-2017, China has spent almost 30 years brewing, piloting, implementing and perfecting the residential electricity tiered pricing. In this paper, we review the four stages of China's tiered electricity price policy, and use data of residential electricity consumption and existing literature on this policy to evaluate its effect.

2. Electricity Price Policies before 2004

2.1. Shortage of Electricity Supply during 1987-1997

The electricity industry has developed quickly since 1949. Total installed capacity of power increased significantly from 1.85 million kilowatts in 1949 to 94 million kilowatts in 1986, and the electricity supply has also reached 449 billion kilowatts in 1986, which ranked fifth in the world, only after the United States, the former Soviet Union, Japan and

Canada. However, per capita electricity consumption in China was only about 4% of the per capita consumption in developed countries, much lower than the world average.

However, due to fast economic growth in China, electricity demand has surpassed the supply since 1970s, leading to sustained power shortage. After the reform and opening up, the power shortage has expanded from some regions to the whole nation. Power grids in Northeast China, North China and East China and other power grids with a large proportion of hydropower component, due to lack of hydropower output in the dry season, coupled with the lack of thermal power capacity, almost cut power every year. By 1986, China's electricity shortage reached about 70 billion kWh. In order to first ensure the living standards of residents, the limited power supply resulted in part of the enterprise production at a state of "stop three days operate four days" and some new enterprises unable to put into production due to lack of electricity. According to Cai and Shen (1987), the direct economic losses caused by the lack of electricity supply in 1986 were more than 40 billion yuan (current price).

There are three reasons for the shortage of power supply:

First, China has underestimated the role of electricity power in economic development. Experience from other countries tells that electricity goes first in the initial and middle stage of industrialization, and the elasticity of electricity (the ratio of electricity growth rate and economic growth rate) should be greater than 1. After reforming and opening up in 1978, high electricity consumption companies (steel, raw materials processing, mechanical manufacturing, chemical) in the light and heavy industries have grown very fast, which leads to a huge demand of electricity. However, during the same period, the grid construction in China was relatively lagging behind, and it relied heavily on the public investment. According to the statistics, despite during the sixth Five-Year Plan the country's new installed power capacity was 20 million kilowatts, during the same period economic growth in China was 10.5% while the power supply growth was only 6.5%, resulting the power elasticity coefficient at only 0.6, well below 1. This is also the root cause of electricity shortage.

Second, after reforming and opening up, together with the income increase, the living standard of Chinese people has improved significantly. Households begin to use all sorts of electrical appliances, leading to a sharp increase of electricity demand. Taking Shanghai as an example, residential electricity consumption in Shanghai has increased at an annual rate of 15% on average during the sixth Five-Year Plan, much more than the growth rate of power supply at the same period.

Third, the electricity price was determined by the government and very cheap. Chinese government took electricity as a planned product rather than commercial goods. The governments (the planning committee of the provincial government) decided the electricity price, in order to make up the production cost rather than make profit. Therefore, it was a subsidy electricity pricing policy. Before 1990, China used the unified pricing policy by province, i.e., provincial governments decided the unified electricity price in their individual province based on the cost of electricity production. In fact, the provincial electricity price had not changed much for three decades till 1990. During the period between 1978 and 1990, to encourage industrial development, China had used more preferential subsidies for the industrial electricity consumption, and thus the residential electricity price was slightly higher than the industrial price in this period. Nevertheless, the residential electricity price was still very cheap, due to the pricing strategy of making up the cost.

During this period, the continued shortage of electricity and priority to ensure the residents of electricity consumption led to expanding losses of Chinese enterprises, which seriously affected China's economic development. Therefore, the Chinese government was determined to solve the problem of power shortage as soon as possible. In addition to the substantial increase in investment in the power industry, the government also began to consider the electricity price reform to promote electricity saving. At that time, there were two proposals for the electricity price reform. First, increasing the price of residential electricity or adopting tiered price policy to increase the cost of residential electricity consumption so as to encourage residents to save electricity; second, increasing the price of industrial electricity to encourage firms to save electricity.

However, due to the move to encourage industrial development during the reform and opening up period, industrial electricity had a larger subsidy, and residential electricity price had already been higher than industrial electricity price. To continuously increase the residential electricity price is clearly not in line with the income level of the residents. In addition, the proportion of residential electricity consumption was still very low (residential electricity accounted for less than 10% of the total electricity consumption). Therefore, even if a small increase in residential electricity price, the effect of alleviating the power shortage would not be obvious. Based on the above considerations, since 1990, China has repeatedly increased industrial electricity price, while residential electricity price had basically not changed. By 2001, in most provinces and cities industrial electricity price has been higher than residential electricity price. From the point of view of relative industrial and residential electricity price, China's electricity price policy has actually been shifted from residential electricity subsidizing industrial electricity to high industrial electricity price subsidizing residential electricity price.

During this period, the Chinese government has begun to explore the feasibility of adopting tiered electricity price to guide residents to save electricity, but only in a small range of discussion and through the learning of experience from abroad. For example, the former state power company in China set up a special demand side management guidance center in December 1997, and the center's main task was to prepare development of a new price policy for the future and conduct all sorts of policy research. After the establishment of the center, they began to go abroad frequently to study the electricity reform experience from other countries, and invited foreign experts to exchange ideas as well. In addition, the center founded the journal of *Demand Side Management* in October 1999, which is so far the only academic journal devoted to the issue of electricity prices in China. For example, in 2000, the journal has a special issue to introduce the lesson China's Henan Electric Power Company hear learnt from Hong

Kong's experience in the implementation of the five-tier electricity price program (Liao and Bai, 2000).

2.2. Excess Electricity Supply during 1998-2003

There was a short period of excess supply of electricity in some provinces during 1998-2003. There are a few reasons: On the one hand, through more than 10 years of development of power industry, the growth of China's electricity supply is relatively fast. For example, during the eighth Five-Year Plan (1991-1995), the installed capacity of electricity has increased by 75 million kilowatts. In line with the economic development, the increase in electricity supply has eased the large-scale electricity shortage. On the other hand, from the demand side, because of the 1998 Asian Financial Crisis, China's domestic economic tightening and some other reasons, a large number of export companies reduced their production, and the new business also delayed their operation due to the pessimistic expectation, which led to significant decline in China's power demand. The utilization rate of many power plants was only about 50%. Changes in both the demand and supply sides of the power industry in this period led to a surplus of electricity in some parts of China or an obvious peak-valley phenomenon in the usage of electricity.

In order to consume the excess supply, the Chinese government had used new preferential policies for the industrial electricity, and local governments also encouraged households to consume more electricity. The policies to encourage residents to consume electricity include two: one is the decreasing tiered pricing strategy and the other is the peak-valley price policy.

2.2.1. Decreasing Tiered Pricing Strategy

In order to encourage electricity consumption, starting from 2000, some regions began to use decreasing tiered pricing strategy. Unlike the usual tiered pricing, decreasing tiered pricing is to price low for more consumption, so as to encourage more consumption. For example, Harbin in Heilongjiang has used this pricing in a period when there was electricity surplus. In order to get the expected effect, Harbin Electric Power Bureau had taken preferential measures for the smooth implementation of the decreasing tiered price, such as free charge for residents who apply for capacity expansion.

Similarly, from 2002 onwards, there has been sustained power supply surplus in energy-rich province Gansu. A large number of power equipment was left idle. In order to encourage residents to use more electricity, Province Electric Power Company in Gansu began to implement the decreasing tiered price policy for urban and rural residents from January 2002. The specific policy is: the first tier has the household monthly electricity consumption of 50 kWh as the base, and the current residential electricity price applied; the second tier is household electricity consumption exceeding 50 kWh and less than 100 kWh, and the unit price was 0.05 yuan cheaper; the third tier is the household monthly electricity consumption between 100 and 150 kWh and the unit price was 0.1 yuan cheaper; the fourth tier has the household electricity consumption of more than 151 kWh, and there was a concession of 0.15 yuan/kWh.

2.2.2. Peak-Valley Price Policy

Another characteristic of this period of excess electricity is also the peak and valley difference in the electricity consumption. Therefore some areas have taken a peak-valley price policy. For example, from September 2001 Zhejiang has piloted the peak-valley price policy. After the pilot, from 20 December 2002, Hangzhou in Zhejiang was the first to formally implement the policy. Residential electricity price was no longer unified as has been used for several decades. It would be adjusted in accordance with the different hours of the day. According to the policy in Hangzhou, the peak hours of electricity consumption was from 4:00 to 22:00, and the peak price was 0.56 yuan/kWh; the valley period was 22:00 to 4:00 the next day, and the valley price was 0.28 yuan/kWh. In addition, in order to accurately measure the electricity consumption, Hangzhou residents could apply to install a new meter on a voluntary basis. The cost to install this meter was about 300 yuan per household, and Hangzhou Electric Power Department bear 2/3 of the installation cost. Since April 2003 the peak-valley price was applied to the whole Zhejiang province, but urban residents can choose to implement it voluntarily. As of June 2008, a total of 1.888 million households implemented the peak-valley electricity price, of which 1.5864 million were urban households, accounting for 26.72% of total urban households; 301.6 thousand were rural households, accounting for 2.7% of total rural households. The consumption during the implementation of the peak-valley electricity price reached 2.09 billion kWh, accounting for 17.6% of the total electricity consumption; and the peak and valley power use ratio was 58:42.

The implementation of the peak-valley price in Zhejiang lowered the average bill of residential electricity consumption. According to statistics in Zhejiang, the effective price for households who have implemented the peak-valley price was about 8 cents/kWh lower than those who did not implement the pricing, and therefore the benefits were very obvious. In the vast rural areas the implementation of the peak-valley price can encourage the majority of rural residents to consume the valley electricity and reduce their burden of electricity cost, so that farmers could also enjoy the quality of electricity services. The implementation of peak-valley price policy is in fact to reduce the peak and fill the valley, and to some extent, played a role to reduce the investment on power construction and save energy. It also played a positive role in easing the province's peak usage of electricity.

But the policy also increased the burden on the power companies. First, the implementation of peak-valley electricity price increased the cost of power grid companies and reduced their revenue. Statistics show that, in 2007 due to the implementation of the peak-valley electricity price the company received 221 million yuan less; in the first half of 2008 the company received 112 million yuan less. Second, it increased the difficulty of power grid operation. At present, Zhejiang implements both the peak-valley price and tiered price, so the grid companies have an increasing workload in the meter reading, meter management, system maintenance and so on.

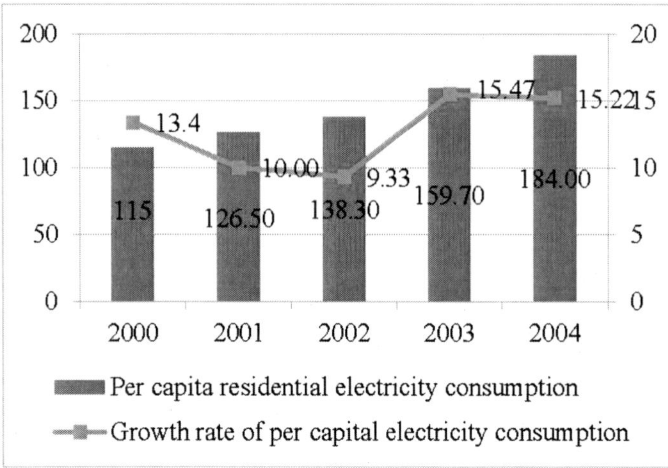

Figure 1. Per-capita residential electricity consumption and its growth rate in China (2000-2004).

After the policies encouraging residents to use electricity, there was a substantial increase in residential electricity consumption. According to the national grid statistics, in 2000 China's per capita residential electricity consumption was only 115 kWh; in 2001 China's per capita residential electricity consumption increased by 10% to 126.5 kWh, and it increased to 138.3 kWh in 2002, an increase of 9.33% from the previous year. The policy of encouraging electricity began to show results in 2003 when per capita residential electricity consumption increased significantly by 15.77% to 159.70 kWh and in 2004 there was another increase of 15.22% to 184.00 kWh.

3. China's Large-Scale Power Shortage of 2004

3.1. Large-Scale Power Shortage in 2004 and Residential Electricity Price Reform

After a brief electricity oversupply, China has experienced an unprecedented large-scale power shortage in 2004, which amounts to a shortage of 20 million kWh that is equivalent to the installed capacity of the Three Gorges Hydropower Station. One of the reasons for the large-scale power shortage is the sustained fast economic growth in China since 2001 and a new round of economic restructure. Many firms in the heavy chemical industry, especially those with high electricity consumption, have grown quickly. In addition, due to the improvement of living standards and earlier policies encouraging electricity consumption, residential electricity consumption has risen quickly. It grows at an annual rate of 15% in 2003 and 2004, and even 20% in 2005, which accounts for 10% of total electricity consumption for the first time. Overall, the significant growth in electricity demand leads to a sharp turn of short oversupply in 2002-2003 to the large-scale shortage in 2004.

There were 24 provinces who had power cuts during January to April 2004; and the most serious shortage of power emerged in the fastest growing Yangtze River Delta region, seriously affecting China's economic development. The annual electricity gap was more than 20 million kWh in 2004, of which Jiangsu alone had a power shortage reaching 8.5 million kWh. In order to ensure the electricity for basic living demand, a large number of enterprises were forced to take "open five stop two" or even "open four stop three" policy. In the peak electricity usage summer of 2004, organizations, institutions, hotels, offices, shopping malls and other units in Nanjing, Jiangsu were forced to use 70% of the normal load of electricity; during the electricity peak (8:30 - 22: 30) the equipment and capacities were specified to stop working to shift the peak; all the city lighting stopped working, including some residential lighting in buildings' staircases. According to the "2004 China Electric Power Industry Annual Report", it was expected China's power gap would still reach 2000-2500 million kWh in 2005.

In the face of severe gaps in power supply and demand, the "power cuts" is only a provisional measure. In fact the economic cost caused by power shortage still exists. To encourage the power industry to increase production capacity requires a longer period, which cannot quickly alleviate the lack of electricity. Therefore, the most effective measure in a short run is to change producers' and consumers' behavior by implementing electricity price reforms. Besides a moderate increase of industrial electricity price, the government gradually reformed the residential electricity price policy, so as to raise the cost and make residents more conservative in electricity consumption.

In 2004, National Development and Reform Commission issued a special report "Interim Measures on The electricity price, transmission and distribution price, and sales price management", and it was formally implemented on 1 May 2005. The Measures clearly require that through the reform of the electricity price under the planned system, establish a sound and reasonable sales price mechanism, and make full use of price leverage to reduce the power gap. It plans to introduce the "Renewable Energy Law" in 2006, and conduct the price and power production reform in the legislative form.

However, due to the widespread impact of electricity price reform on all the residents, and the vast differences in the development of various regions, the reform may be resisted if suddenly implemented across the whole country. Therefore, the Measures also stated that "the residential and agricultural production price, based on the average price of various voltage levels, taking into account the users' capacity, maintain at a relatively stable level. If residential and agricultural production price is lower than the average price, the spread is shared by the business and other users."

3.2. Policy Evaluations in the Three Pilot Provinces

Therefore, for the sake of prudence, the Chinese government decided to select some provinces to carry out the pilot work of tiered pricing. The selection criteria are: the residents are relatively wealthy and able to afford the change; there is serious shortage of power in the province. Zhejiang and Fujian are provinces that have a high level of economic development; residents in these two provinces have a high level of tolerance; they were frequently affected by the large-scale power cut in 2004 and thus have a profound understanding of the importance of electricity saving. In addition, Sichuan is the most populous province and used as a representative of the less developed central and western provinces. Given the above considerations, during the period of 2004-2006, the Chinese government has selected these three provinces - Zhejiang, Fujian and Sichuan, to carry out the pilot study of the residential tiered electricity pricing reform, with the purpose to accumulate experience for the full implementation of the policy in the future. The specifics of the pilot are introduced below:

3.2.1. Pilot Study: Zhejiang

Before the implementation of the tiered price, the electricity pricing policy in Zhejiang has been preferential to the residents. Even in the time of continuously rising power costs, the residential electricity price has basically maintained its stability, resulting in a serious deviation from power supply cost. Under the unified low-cost policy, the high and low-income groups seem to have enjoyed the same subsidy. However, because the low-income groups consume much less than the high-income groups, it in fact results in high-income groups getting more subsidies than low-income groups. This is called the "Matthew effect".

In addition, energy resources are not rich in the economically developed province. More than 95% of primary energy are imported from other places. In 2003-2004, Zhejiang suffered a serious power shortage; it had the largest scale of power cut nationwide. To promote a conservation-oriented society, and guide residents to use electricity rationally and conservatively, considering the level of economic and social development, power supply capacity and demand, the ability of the masses and the electricity price in the neighboring provinces and cities, Zhejiang in 2004 put forward the first proposal to implement tiered progressive electricity price. Through the hearing, and approved by the National Development and Reform Commission, it was formally implemented starting from 1 August 2004.

At that time, the specific of the tiered price in Zhejiang is to change the unified electricity price in the past, and divide the monthly household electricity consumption into three tiers. The first tier is household electricity consumption of no more than 50 kWh and the electricity price of 0.53 yuan/kWh; the second tier is between 51 kWh and 200 kWh and the unit price increase 0.03 yuan to 0.56 yuan/kWh; the third tier is more than 200 kWh and the additional unit price increase of 0.10 yuan to 0.66 yuan/kWh.

After the implementation of the policy, Zhejiang Electric Power Corporation continuously tracked the household electricity consumption and found that before the tiered pricing implementation, residential electricity consumption increased by 17.04% and 24.2% in 2002 and 2003; after the implementation, except in 2004 due to power cuts, residential electricity consumption fell by 6.47%, which results in a large increase of 24.35% in 2005, residential electricity consumption increased by 16.24%, 15.13% and 14.40% in 2006, 2007 and 2009, respectively. The annual increase rate is lower than those before the implementation, and also lower than the national average growth rate.

Furthermore, in order to examine whether the tiered price increased the burden on low- and middle-income groups, Zhejiang Electric Power Corporation also conducted a micro-level multi-year survey on the electricity consumption of residents in Zhejiang. The track data show that 86.7% of the households belonged to the first tier and the slightly fare-increase second tier, so the impact on the electricity bill is small. As of 2007, 16.7 million households in both urban and rural Zhejiang have the electricity meter, accounting for 98.42% of the total households. Among them, 5.63 million are urban households, and 11.07 million are rural. Their 2007 total annual electricity consumption was about 20.569 billion kWh, and the first-tier electricity use accounted for 33.49%; the second-tier electricity use accounted for 39.58%; and third-tier electricity use accounted for the rest 26.93%. The average monthly electricity consumption per household was about 102.59 kWh; it is on average 172.47 kWh per month in urban, and 67.04 kWh in rural. Households with the average monthly electricity use at 50 kWh and below accounted for 44.7%; households with the average monthly electricity use between 51 to 200 kWh accounted for 42.0%; and the remaining 13.3% households consumed more than 201 kWh per month (Jago, 2008).

In addition, in order to investigate the policy impact in cities with the middle-income level, Zhejiang Electric Power Corporation conducted the follow-up survey on the monthly electricity consumption for residents in Yongkang city in Zhejiang. The results showed that after the implementation of the tiered price since 1 August 2004, 68% of the residents did not pay larger electricity bill due to the price increase. Results from the micro-survey in January-June 2006 found that the average price increase was 0.032 yuan/kWh, 0.77 cents lower than the price published in the hearing. Specifically, in the first half of 2006, the total electricity consumption in Yongkang was 83.17 million, the users were 215.6 thousand, the average monthly electricity consumption per household was 64.3 kWh, of which 147 thousand households (68.18%) had monthly electricity use less than 50 kWh, and the excess charge thus had no impact on them. The impact on rural households was smaller. For example, the village of Zhaikou in Yongkang city (with medium level of per capita income in Yongkang) had a total of 330 households. 304 households had the electricity consumption within 50 kWh in June, which means that 92% of the households were not affected by the tiered price (Shi, 2006).

After four years' implementation of the tiered electricity price, Zhejiang also pointed out the problems to be solved in the future. First, the legibility and rationality of the "household". At present, there is no clear legal definition of the household; dispute exists on whether the household is defined by the property certificate or by the public security registration account. Second, with the economic development and improvement of living standards, the tier setting should be timely adjusted. The current first tier electricity consumption cannot meet the basic needs of residents, and therefore the government should make timely adjustment. Third, the rationality of the price increase across tiers. There is a lack of scientific support for how the tiered pricing should be set. That the extent to which the price should increase can encourage households to save electricity is difficult to estimate, and different groups of people have different opinions. Forth, the attribution of the additional income from the tariff increase. Who and where does the additional income go is hard to explain to the hearing representatives.

3.2.2. Pilot Study: Fujian

Fujian is the second province to join the pilot study and it officially launched the tiered electricity price on 1 October 2004. As the income gap between urban and rural residents in Fujian was large, and there were a large number of households who shared the usage of the meter, the tiered price was implemented only for nine cities, in which there were about 3.8 million households who had the meter and used the electricity directly from the provincial power grid. Its specific policy is similar to that in Zhejiang. It has three tiers: the monthly household electricity consumption at 150 kWh and below is charged a price of 0.4463 yuan per kWh; monthly consumption between 151 and 400 degrees is charged a price of 0.4663 yuan per kWh (2 cent increase); monthly consumption at 400 degrees and above is charged 0.4463+0.12 yuan per kWh. For households who shared the meter, the unified price of 0.4643 yuan per kWh, i.e., the second-tier price, applied.

Residential electricity consumption in Fujian increased by 23% in 204, but in 2005 and 2006, the increase dropped to 16% and 15%, respectively. The reasons are as follows. First, the first-tier was set at a very high level of 150 kWh per month. Most of the households did not exceed the first-tier. Second, the fare increase across tiers is very small, resulting in insignificant bill increase. According to the tiered price in Fujian, if a household consumes over 400 kWh in a month, the electricity bill would increase only 5 yuan due to the tiered pricing. The spread of the prices between tiers is very small, and therefore it basically had no impact on residents' electricity-consuming behavior. People's Daily reporter conducted a small-scale survey in 2009, and the results found that most of the residents were not very sensitive to the tiered electricity price policy due to the small proportion of the additional expenditure (Wei, 2009).

3.2.3. Pilot Study: Sichuan

After two years of pilot work in Zhejiang and Fujian, the State Council decided to have another pilot work in the most populous province: Sichuan, so as to evaluate the policy effect in the lower-income level central and western regions.

It is worth noting that, in Zhejiang and Fujian, the ladder price pilot process, due to the absence of a large-scale price hearing in advance, the tiered price reform resulted in some residents unhappy with not fully involved in the process of the price setting, and some residents unaware of the policy change. Therefore, in order to get public participation, the National Development and Reform Commission commissioned the Sichuan Provincial Price Bureau to hold a major price reform hearing in December 2005, which is China's first public hearing of the tiered-price reform. The contents of the hearing mainly included residential electricity price adjustment and immigration fund co-ordination. The residential electricity price adjustment got a widespread attention due to its relevance to the tiered-price reform (He, 2005a). There were 30 hearing representatives and more than 10 attendees who had participated in the meeting held in the Jinniu Hotel, and representatives discussed and argued the feasibility and necessity of the price adjustment. At the hearing, about 70% of the representatives were in favor of the policy, but they also made a lot of suggestions. The price department in Sichuan collected all the views and submitted them to the National Development and Reform Commission after the hearing (He, 2005b).

Since 30 June 2006, Sichuan implemented the tiered electricity price for urban and rural households with electricity meters. From the policy point of view, the tiered price policy in Sichuan is quite different from that in Zhejiang and Fujian. The policy states the residential electricity consumption in Sichuan is divided into four tiers: the first tier is the monthly electricity consumption at 60 kWh and below, and the price is maintained at 0.4724 yuan/kWh; The second tier is the monthly electricity consumption between 61 and 100 kWh and the electricity price for this part increases 0.08 yuan to 0.5524 yuan/kWh; the third tier is the monthly electricity consumption between 101 and 150 kWh and the electricity price for this tier increases 0.11 yuan to 0.5824 yuan/kWh; and the fourth tier is the monthly electricity consumption of more than 151 kWh and the price increases 0.16 yuan to 0.6324 yuan/kWh. For households who share one meter, the price increases 0.05 yuan, i.e., 0.5224 yuan/kWh. It can be seen that the tiered electricity price in Sichuan is not only divided into more complex four tiers, and the fare increase across tiers is relatively large; in addition, the starting point of its first tier is relatively low at only 60 kWh.

In addition, besides encouraging residents to use the tiered electricity price, Sichuan also developed peak and valley time-tariffs to encourage residents to use electricity at the valley time. It was determined the daily 7 am to 11 am and 7 pm to 11 pm as the peak time, and 11 pm to 7 am the next day as the valley period. In the wet period (June-October) valley time, the unit price would drop from the benchmark price to 0.151 yuan; in the dry-flat water period (November - the following May) valley time, the unit price would drop from the benchmark price to 0.2295 yuan.

After the implementation of the policy, the residential electricity consumption dropped rapidly. Before 2006, residential electricity consumption grew at an average annual rate of more than 12.6% and even reached 18% in 2005. In 2007 when the tiered price was implemented, the growth rate of residential electricity consumption in Sichuan was only 0.11% (almost no growth), and in 2008 the growth rate was 8.95%. Using the lowest growth rate before 2006, it was estimated that Sichuan had saved 1.16 billion kWh in 2007 and 800 million kWh in 2008.

In addition, the tiered pricing policy also led residents to purchase energy-saving appliances. A survey in Chengdu shows that after 2006 the tiered price was implemented, the proportion of people who bought energy-saving lamps increased significantly. Of the 50 households surveyed, 34 households used energy-saving lamps, accounting for nearly 70% of the sample. The phenomena that new homes use energy-saving lamps is particularly prominent. Of the 50 households, 21 households moved to the new house in the last two years and 19 of them used energy-saving lighting (Luo, 2007). The sales data in Sichuan NVC Lighting Co., Ltd show that 90% of the people who buy lamps will choose energy-saving lamps. It is estimated that, since heating, air conditioning and lighting accounted for 14% of the total energy consumption, if all households in Sichuan replaced incandescent lamps with energy-saving lamps, energy saved in two days in Chengdu would be enough to supply for lighting for one day.

However, it is generally reflected by the people in Sichuan that 60 kWh baseline is too low, resulting in the vast majority of users must pay an extra charge. According to the big data on electricity consumption, the average monthly electricity consumption in Chengdu in 2008 was about 82 million kWh, with only 23% users with electricity consumption of 60 kWh or less, 29% users of 61- 100 kWh; 22% users of 101 -150 kWh, and up to 26% of users with more than the upper limit of 151 degrees. Compared to "the coverage of the first tier of electricity consumption is 70% of households", a guidance proposed by the National Development and Reform Commission, the coverage of the first-tier in Sichuan is apparently too low.

3.3. Evaluation on the Pilot Work in the Three Provinces

Although the pilot work of the tiered price in the three provinces has achieved some success, there are some urgent problems to be solved, which are discussed below:

1. The determination of the benchmark of the first tier. The first tier of electricity consumption is one of the most concerned problems for many people. This is because the first tier is designed for low-income groups' use of electricity, and it should meet the electricity needs of minimum living standards. In the context that the residential electricity price is generally low, if the first tier is set too high, it is equivalent to continuously implement the past low price policy. In Sichuan, the first tier of electricity was set at 60 kWh. This is considered relatively high, because in 2008, the average annual electricity consumption per household in Sichuan was 778 kWh, and the average monthly electricity consumption per household was 64.8 kWh, while the first tier of electricity consumption was close to the average. The first tier in Zhejiang was the lowest among the three provinces. The average annual electricity consumption per household in Zhejiang was 1506 kWh,

and the average monthly electricity consumption per household was 125.5 kWh, indicating that the first tier set at 50 kWh was appropriate. In Fujian, the first-tier electricity consumption was set at 150 kWh, obviously too high. In 2008, The average monthly electricity consumption per household was 131 kWh, while the first tier was set much higher than the average level.

2. The price difference between the tiers. The key to implementing the tiered price is the price gap between the tiers, especially between the first tier and the second tier. Usually the first-tier price is to take care of low-income groups; thus it is in principle subsidized electricity prices. The second-tier price should be a reasonable normal price, and prices for the third and the subsequent tiers should be gradually higher. From the three pilot provinces, price differences for the first tier and the second tier were only a few cents. In particular, in Fujian, due to its first-tier electricity use is set too high, and the price gap between the first and second tiers is not large, the residents in Fujian were not very sensitive to the tiered price, and eventually the electricity saving effect was not obvious.

From the perspective of the actual implementation, the tiered price reform was implemented more smoothly in Zhejiang and Fujian. In Sichuan, there was much resistance, and the community had more complaints. The main reason is that the tiered price setting in Fujian is in line with the wishes of the residents and the starting point of the tiers is set high, which ensure that electricity bills of the majority residents' were not raised or only fine-tuned. The level of economic development in Zhejiang is high and the residents are not sensitive to the price adjustment. However, in Sichuan, not only too many tiers were set, the price increase was too much, and the first-tier coverage was too low, resulting in the majority of residents must pay an additional fee. Therefore more complaints were received in Sichuan.

The greatest experience from the pilot electricity price reforms in the three provinces is that the first-tier should be positioned to meet the needs of low-income groups or meet the needs of the vast majority of household electricity demand. If the former, the policy effect will be obvious, but most families need to increase the electricity bill. Therefore, it may easily lead to households, especially those in lower income groups to oppose it; if the latter, since most families do not need to pay extra, the policy is easier to implement, but it also means that the policy effect will not be too obvious.

4. Electricity Price Policies between 2008-2010

After the 2004-2008 tiered pricing in three provinces at the pilot stage, due to the positive effect on electricity saving and a power gap occurring again in 2008, the National Development and Reform Commission began to prepare for the nationwide implementation of the tiered pricing. However, due to some obstacles, the policy was not implemented during 2008-2010. The main reasons are as follows:

1. Technical reasons. On the one hand, China's coverage of "one household one electric meter" was too low. In 2008, the coverage rate of the electric meter was only 60%, and there were a large number of users who shared the same meter. These lead to the difficulty in reasonably determining the basic electricity consumption for each household. In order to implement the progressive fare increase, every household need to have an individual electric meter; this would cost a considerable amount of money and time. On the other hand, even in the household with the meter, the old meters were not able to record and calculate the tiered tariff and use the prepaid card. It would be another huge investment and also take a long time to upgrade the devices to allow for remote electronic meter reading; if taking manual method, it would not only increase the cost of electricity charges, but also a technical regression.

2. During this period electricity price has been raised for a few times, and people may become dissatisfied with another change to tiered pricing. Starting from the large-scale power shortage in 2004, China has raised its electricity price almost every year, and the rate of increase is not trivial. In 2006, China's average electricity price was 0.443 yuan/kWh; it increased to 0.445 yuan/kWh in 2007 and 0.497 yuan/kWh in 2008, which amounts to an increase of 12% within three years. The continuous increase in electricity price was seen in almost every province. So tiered pricing reform may lead to dissatisfaction from low-income families.
3. There was debate on how to determine the first-tier price, and "87 kWh red line" was denied by the hearing. The price department in many provinces began to hold the hearing on tiered pricing in 2009; it turned out that all the hearing results vetoed the tiered pricing, especially that about the first-tier setting at only 87 kWh per month. Hence, the tiered pricing was basically shelved. Challenged by the results of the national hearing, Mr. Cao Changqing from the pricing division of the National Development and Reform Commission said On 22 November 2009, the monthly electricity demand for basic living needs is to be determined by the individual province based on the local conditions, the state will not develop unified standards. Cao Changqing said in an exclusive interview, the 87 kWh was the average monthly electricity consumption in a household provided by the National Bureau of Statistics of China. Because of the vast difference in natural resource, living habits, and economic development across regions, the state will not use the "one size fits all" method to determine the unified electricity consumption for the basic living needs across the whole country. The denial of the "87 kWh red line" marked that the tiered pricing is unlikely to launch in a short term. A large number of reports and articles criticizing the tiered pricing appearing in 2009 also reflects the situation.

During this period, the Chinese government was still preparing for the implementation of the tiered pricing reform in the future. The first step was to improve the ability to collect accurate household electricity consumption data. In order to solve the technical barrier of the tiered pricing, on 25 September 2008, the State Grid Corporation held a seminar on the construction of power information collection system for the low-voltage users in Tianjin, which indicated the beginning of the technical preparation for the implementation of future tiered pricing (Lu, 2008). The meeting was held when the country aimed to achieve the goal of energy saving and emission reduction. It pointed out that energy-saving and emission reduction will be an important task for a long time, and tiered pricing is one of the means to promote energy-saving and emission reduction; centralized meter reading will be one of the requirements for modern management of the power grid. Staring from 2009, the large-scale of power information collection for low-voltage users launched, which means that the technical preparation work for the tiered pricing reform has never stopped, even though it would not be introduced soon.

China continued the preparation for the implementation of tiered pricing. In October 2010, the Development and Reform Commission issued the "Opinion on the Implementation of Tiered Pricing for Residential Electricity" (Opinion thereafter), and received a total of 21,794 feedback through the Internet, fax and letters. Among them, 61% expressed support for the tiered pricing design and the direction of reform, and made suggestions and comments from different perspectives.

There were two proposals in the Opinion. Proposal 1: the first tier covers 70% of the household average monthly electricity consumption. Preliminary statistics show that households who consume 110 kWh or less per month account for 70% of the total number of households. The first-tier price maintains the basic electricity price. The second tier is determined by the consumption which can cover 90% of the household average monthly electricity consumption and the national average at this level is 210 kWh. The price increase should be not less than 5 cents/kWh.

The third tier is more than 210 kWh, and the price increase compared to the baseline should be not less than 20 cents/kWh.

Proposal 2: The first tier is determined at the national average of 140 kWh which covered 80% of the household average monthly electricity consumption, the price was raised by 1 cents/kWh from the initial stage. The second tier is decided by the consumption that covers 95% of the residents, which was 270 kWh. The price increase should be not less than 5 cents/kWh. The third tier consumes more than 270 kWh, and the price increase should not be less than 20 cents/kWh.

5. Electricity Price Policies in 2012

5.1. Final Preparation for the Implementation of the Tiered Electricity Price

In 2011, rising coal prices, fast-growing energy-intensive industries, and the drought in the south altogether led to a substantial decline in hydropower production and a power shortage of 30 million kWh. It was the largest scale of electricity shortage after 2004. Many provinces cut power frequently, and people's life was seriously affected. This has provided sufficient reasons for the implementation of the tiered pricing. On 8 June 2011, in the context of the large-scale power shortage, the person in charge of a department in the National Development and Reform Commission told reporters, in an interview of "2011 Key Work for Deepening the Reform of the Economic System", that specific implementation of the tiered pricing will be carried out soon in the coming 2012.

Results from the "Survey on residential electricity consumption and tiered pricing policy" released by the Friends of Nature, an environmental protection organization, in June 2011 show that most residents realized the short supply of city power, and supported the adoption of tiered pricing so as to save energy. Among them, people in Beijing showed the highest rate of support (73.6%), which was followed by Xiamen (69.9%), and subsequently Zhengzhou (65.9%).

In May 2012, Chinese provinces held intensive hearings on the implementation of the program so as to collect opinions from the society and further improve the program. There were 29 provinces who had held a hearing, and nearly a thousand participants till early May. In June 2012, the National Development and Reform Commission announced in the tiered pricing press conference that 29 provinces and municipalities (except Xinjiang and Tibet) had completed the hearing, and would implement the tiered pricing gradually; starting from 1 July 2012, the policy would be implemented nationwide. Since China has a long history of government subsidizing residential electricity use for more than 30 years, the proposed tiered pricing is a correction to the long-term low price, and reflects the reasonable cost of electricity.

The National Development and Reform Commission said that according to the current policy design, a majority of regions would have the first block covering more than 80% of electricity consumption, and this number may go beyond 90% in some regions.

5.2. Characteristics of the Tiered Electricity Price in 2012

As the 2012 tiered price policy is made by individual provinces within the general principle, the policies are not the same across provinces; in some regions the tiered price policy is even introduced on top of other existing tariff policies. So the details are quite complex. Table 1 shows the specifics of the tiered pricing scheme of 29 provinces, municipalities and autonomous regions in China on 1 July 2012, including the pricing method, the number of tiers and the range of the electricity used in each province, the basic price and other electricity tariff polies.

Table 1. Specifics of the tiered price policy in 2012

Province	Pricing method	First-tier	Second-tier	Third-tier	Electricity price (yuan/kWh)
Beijing	Yearly	0 - 240	241 — 400	>400	0.4883
	Peak-valley and sharing-meter users	Continue the current peak-valley price			
Shanghai	Yearly	0 - 260	260-400	>400	0.617
		Peak-valley: no adjustment	**Normal:** +0.05yuan/kWh; **Peak-valley:** +0.06yuan/kWh at peak, +0.03yuan/kWh at valley	**Normal:** +0.3yuan/kWh; **Peak-valley:** +0.36yuan/kWh at peak, +0.18yuan/kWh at valley	
Tianjin	Yearly	0 - 220	221 - 400	>400	0.49
Zhejiang	Yearly	0 - 230	231 - 400	>400	0.538
Fujian	Monthly	0 - 200	201 - 400	>400	0.4983
	Alternative: **Peak:** 200 kWh or below, 0.5283 yuan; 201-400 kWh, 0.5783yuan, 401 kWh or above, 0.8283 yuan **Valley:** 200 kWh or below, 0.2983 yuan; 201-400 kWh, 0.3483yuan, 401 kWh or above, 0.5983 yuan **Sharing-meter users:** +0.035 yuan/kWh				
Shandong	Yearly	0 - 210	211-400	>400	0.5469
Guangdong	Summer (May-Oct.)	0 - 260	261 - 600	>600	0.61
	Non-summer (Nov.-Apr.)	0 - 200	201 - 400	>400	
Jiangsu	Yearly	0 - 230	231 - 400	>400	0.5283
	Peak-valley	Peak: 0.5583 yuan/kWh, Valley: 0.3583 yuan/kWh			
Hubei	Monthly	0 - 180	181-400	>400	0.57
Sichuan	Monthly	0 - 180	181-280	>280	0.5244
	Keep the valley price promotion: rich water (Jun.-Oct.) valley price 0.175 yuan, low and flat water (Nov.-May.) valley price 0.2535 yuan.				
Henan	Yearly	0 - 180	181 - 260	>260	0.56
Chongqing	Monthly	0 - 200	201 - 400	>400	0.52
Jiangxi	Monthly	0 - 180	181 - 350	>350	0.60
	Peak-valley	**Alternative:** Peak: +0.03yuan; Valley: -0.12yuan			
Hunan	Spring Autumn (Mar.-May, Sep-Nov.)	0 - 180	181 - 350	>350	0.588

Province	Pricing method	First-tier	Second-tier	Third-tier	Electricity price (yuan/kWh)
	Summer Winter (Jan., Feb., Jun.-Aug. Dec.)	0 - 180	181 - 45 -	>450	0.588
Hainan	Summer (Apr.-Oct.)	220	221 - 360	>360	0.6083
	Non-summer (Nov.-Mar.)	0 - 160	161 - 290	>290	
Yunnan	Rich water (May-Nov.)	A preferential price of 0.45 yuan			0.483
	Low water (Dec.- Apr.)	0 - 170, 0.45yuan	171 - 260, 0.5yuan	>260, 0.8yuan	0.45
Liaoning	Yearly	0 - 180	181 - 280	>280	0.50
Hebei	Yearly	0 - 180	181 - 280	>280	0.52
Jilin	Yearly	0 - 170	171 - 260	>260	0.525
Shanxi	Monthly	0 - 170	171 - 260	>260	0.477
Guangxi	Peak	0 - 190	191 - 290	>290	0.538
	Non-peak	0 - 150	151 - 250	>250	
Heilongjiang	Yearly	0 - 170	171 - 260	>260	0.51
Gansu	Monthly	0 - 160	171 - 240	>240	0.51
Ningxia	Yearly	0 - 170	171 - 260	>260	0.4486
Qinghai	Yearly	0 - 150	151 - 230	>230	0.4271
Anhui	Yearly	0 - 180	181 - 350	>350	0.56
Guizhou	Winter (Dec.-Mar.)	0 - 210	211 - 380	>380	0.4556
	Non-winter (Apr.-Nov.)	0 - 170	171 - 310	>310	
Inner Mongolia (West)	Yearly	0 - 170	171 - 260	>260	0.43
Inner Mongolia (East)	Yearly	0 - 180	181 - 350	>350	0.5

It can be seen from the table, there are both consistencies and discrepancies in the tiered price schemes across provinces in 2012. This is because China's tiered pricing is a combination of national general standards and local specific standards. There are a few characteristics:

1. The number of tiers and the tiered price setting adopt a unified national standards. All provinces have adopted the unified tier-setting standards and minimum price increase proposed by the

National Development and Reform Commission to determine the tiers and tiered prices. As can be seen from Table 1, the household electricity consumption of the tiered pricing scheme is divided into three tiers in all 29 provinces and autonomous regions. The first tier uses the current basic price, the second tier increases the basic price by 0.05 yuan/kWh, and the third tier uses an increase of 0.3 yuan/kWh from the basic price. Obviously, in setting the number of tiers and the price increase across tiers, all provinces used more or less the same principle. That is to set three tiers and use the national minimum price increase standard.

2. The electricity consumption in every tier is set quite differently. As there is a big difference in the development level across regions, the tiered electricity use is determined by the provinces. In principle, it should ensure that the coverage of the first tier is not less than 80%. It can be seen there is a significant difference in the starting point of the first tier in all 29 provinces. The more developed the economy is, the higher the electricity consumption of the first tier. For example, Shanghai, Beijing, Jiangsu, Zhejiang, Guangdong and other economically developed areas have the first-tier electricity consumption more than 200 kWh, of which Shanghai's first-tier electricity consumption is set to 260 kWh, the highest among all provinces. While the less developed regions such as Henan, Jiangxi, Hubei and others have the first-tier electricity consumption set at a level less than 180 kWh; in Qinghai it is even at 150 kWh. The starting point of the third tier also has a similar situation that in more developed regions, the third-tier starting point is set at more than 400 kWh (in Guangdong, due to the hot summer, the third-tier is even set to more than 600 kWh), while in economically less developed areas, the third-tier starting point is generally less than 300 kWh (Qinghai has the lowest third-tier starting point at 230 kWh, which is even lower than the first-tier consumption in most developed areas). Of course, according to statistics from the National Development and Reform Commission, the vast majority of regions reached the

objective to cover at least 80% of households in the first-tier electricity use, in some areas it is even more than 90%.

3. There is a difference in whether the tiered price is added to the existing tariff policy. As some regions have implemented the peak-valley price or seasonal preferential price before 2012, there are some differences in whether the policy continues after the introduction of the tiered price. In Sichuan and Yunnan that rely heavily on hydroelectric power generation, due to the obvious wet and dry periods, the two provinces adopt the tiered price policy on top of the original seasonal preferential price policy. For Fujian, Shanghai, Beijing, Jiangsu and Jiangxi which has adopted the peak-valley policy, there are many different forms: Shanghai, Fujian and Jiangxi introduced the tiered price on top of the peak-valley price policy, which allows the maximum amount of promotions. Jiangsu allows users to voluntarily choose either peak-valley price or the tiered price, but not the two at the same time. Beijing adopted the tiered price for all households with meters but not previously using the peak-valley pricing; for users who have already chosen the peak-valley price, the tiered price does not apply for the moment.
4. There are differences in whether the tariff is charged on a monthly or yearly basis. Due to existence of differences in seasonal electricity consumption, i.e., electricity consumption is obviously much more in summer for all places and in winter for some places with non-heating, household electricity consumption would be generally exceed the first tier. However, the average monthly electricity consumption may not necessarily be more than the top of the first tier. If using the monthly pricing method, the seasonal factor cannot be ironed. If using the yearly pricing method, households can adjust their electricity consumption within a year, which can ease the seasonal differences in the electricity bills caused by fluctuations in electricity consumption and reduce household electricity costs. For this issue, different regions have adopted different pricing methods. Most regions take into account

the issue, and used the annual average monthly electricity consumption to iron out seasonal factors. Guangdong, Hunan, Hainan, Guangxi and Guizhou and other provinces clearly set the electricity tiers by the peak and non-peak season, so that the first-tier consumption in the peak season is much higher than that in the non-peak season. Only Hubei, Chongqing, Guizhou, Shanxi, and Gansu temporarily adopted a monthly calculation, and did not consider the impact of seasonal factors.

5. Price policy for users who share the meters. Although the coverage rate of the one household one meter has increased tremendously through numerous efforts, by 2012 there was still about 7 million households who used the meter on a sharing basis. For this group of users, they basically continue to implement the past price policy. For example, in Beijing, they still follow the peak-valley tariff policy. Fujian has also taken the similar measure for the meter-sharing users, but the price is increased by 0.035 yuan/kWh.
6. Subsidy policy for low-income households. For electricity subsidies for the low-income groups reflected in the pre-survey, the basic design of the subsidy policy is generally to give each low-income household 10-15 kWh of free electricity. For example, in Gansu and Chongqing the households that are eligible to the "subsistence allowance" and rural "five guarantees" can enjoy 10 kWh of free electricity, and low-income groups in Guangdong can receive 15 kWh of free electricity per month.
7. Electric heating price billing. Beijing, Liaoning, Hebei and Shandong have considered households that use electric heating in winter explicitly. In Beijing the tiered pricing does not apply to households who have shifted from coal to electricity in heating temporarily, and they continue to use the peak-valley electricity price. Liaoning use the sharing-meter price policy during the heating period for households living in communities with central electrical heating (0.02 yuan/kWh increase). Hebei and other provinces have adopted a similar approach. In Shandong, for non-central heating users whose annual electricity consumption is more

than 4800 kWh (third tier), the amount in excess of 4800 kWh is charged in accordance with the second tier which increases only 0.05 yuan, rather than 0.3 yuan increase in the standard implementation for the third-tier use. But there are still some regions where people have to endure the cold so as to save electricity, because there is no "electric heating does not implement the tiered price" provision in these provinces. In addition, in the non-heating areas, although a large number of residents in the winter use electrical heating, there are no corresponding preferential policies.

5.3. Policy Effects

After the implementation of tiered tariffs in 2012, the policy effect has been the focus. However, due to lack of micro-level data available, the literature that have studied the 2012 tiered price reform at the household level is relatively scant. We can therefore only review the effects from the macro and provincial level. From the current available public data, there are at least some effects on energy saving. The effects are discussed below.

5.3.1. The Residential Electricity Consumption in Aggregate and in Per Capita Term has Declined after the Tiered Price Policy

From Figure 2 we can see that the annual growth rate of China's household electricity consumption during 2010-2012 before the implementation of tiered-pricing shows an increasing pattern; it increases from 5.18% in 2010 quickly to 10.66% in 2012. After the implementation, due to extreme weather, the growth rate in 2013 continues to increase slightly. But it drops to 2.67% and 5.42% in 2014 and 2015, respectively. The magnitude of the decline is significant. The same trend is also observed in per capita residential electricity consumption. During 2010-2012, China's per capita electricity consumption increased from 383 kWh to 460 kWh. The growth rate in 2012 is even more than 10%. However, starting from 2014, the per capita growth rate of electricity consumption declined significantly, to only 2.14% in 2014, and no more than 5% in 2015. Taking into account other factors that affect households' electricity consumption during the same period such as per capita income, housing area, electrical ownership and other variables which have showed a steady growth trend, it can be estimated that after 2013, the main reason for the decline in electricity consumption in China is mainly the tiered price.

Growth of residential electricity consumption after the tiered price also slowed down significantly in all regions. Take the economically developed major electricity consuming province – Guangdong as an example. Before the tiered price reform, the annual growth rates of residential electricity consumption are 10.59%, 13.46% and 13.98% in 2010, 2011, and 2012, respectively, and the trend is increasing year by year. After the implementation of the tiered price policy, the residential electricity consumption in the whole province from August 2012 to April 2013 is 52.423 billion kWh; the electricity growth rate is 5.1%, which is 8.36 percentage points lower than before. During August to December 2012, the growth rate is 6.55%, 7.3% lower than that consumed in January-July 2012; during January to April 2013, the growth rate fell further to 2.88%, which is a significant decrease. Annual growth rate of electricity sales is 2.95% in 2013. Compared with 2011, it has decreased by 10.51%.

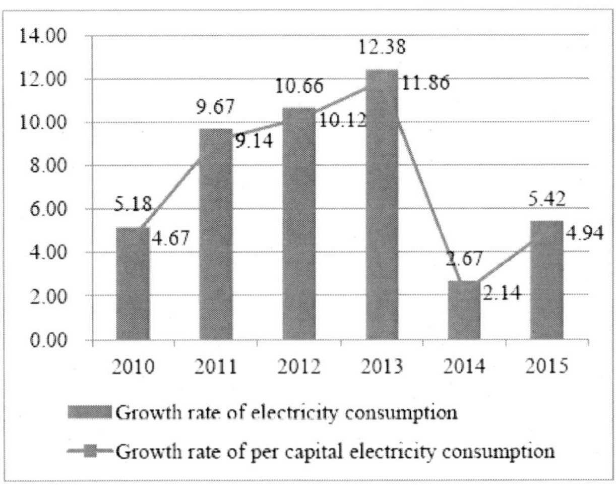

Figure 2. Growth rate of per capital electricity consumption before and after 2012 (%).

In addition, in addition to developed regions, the policy effect in the central and western regions is also obvious. According to data from the Chongqing Electric Power, growth in residential electricity consumption decreased significantly after the tiered price. Looking from the monthly household electricity consumption, the monthly electricity consumption per household in Chongqing is 78 kWh, 6 kWh lower than before. The average household electricity consumption in the first tier increased by 5 kWh, while the average household electricity consumption in the second and third tiers were reduced by 21 kWh and 116 kWh, respectively. Using the average number of households 10.68 million units to estimate, the tiered price policy has saved 770 million kWh of electricity, which is equivalent to 250,000 tons of standard coal, and therefore reduced 650,000 tons of carbon emissions.

5.3.2. After the Implementation of the Tiered Price, Structure of Electricity Consumption has Changed Significantly, and the Share of Households with Large Electricity Consumption Decreased

In July 2013 one year after the implementation of the tiered price, China Power Corporation did a study and found that over 90% of the households in most regions electricity consumption did not exceed the first tier, and the proportion of households falling to the third tier began to decline. The same finding is also observed using regional data. Take Guangdong and Chongqing as an example. Electricity consumption data in Guangdong show that looking from the structure of electricity sales, households who have consumed more electricity are guided to change their behavior by the tiered price and use electricity more rationally. Data showed that in 2013 shares of households who belong to the first and second tiers increased by 4.78% and 0.2%, respectively, while the share of the third-tier household has dropped by 4.97%. That is, 91%, 7.09%, 1.91% of households fall to the first-tier, second-tier, and third-tier electricity consumption, respectively; and the first-tier households have increased to a large scale compared to 2012. From this, we can see there is an increasing awareness of energy saving among residents.

In addition, a small number of regional micro-level studies on the tiered price reform show that tiered price has a significant effect on household electricity consumption, but the effects are different in different income groups. Liang (2013) used the household data of 2000 households in Beijing and the results show that after controlling for other factors, the tiered price led to significant decrease of electricity consumption in all income levels, and the average decline is about 10%. Among them, the decline of household electricity consumption in high-income families and middle and high-income families was the most significant, the decline of household electricity consumption in low- and middle-income households was median, and the decline in in middle-low income levels was not obvious. The results are consistent with the expected target of the tiered price policy, that is, to promote the energy-saving behavior in households that consume more electricity, while affect households that consume less electricity to a minimal level. However, Li (2017) found in a micro-level survey of households in Anhui that although the tiered price policy has certain impact on the residents' electricity-saving behavior, the lowest and the highest income group are not sensitive to the tiered price.

5.3.3. The Power Shortage in China Disappeared, and Sales of Energy-Saving Appliances Increased Significantly

Since 2012, the long-lasting phenomenon of electricity shortage has eased. Except the 2013 summer when continued high temperature lead to short-term local power supply tension, there has been no large-scale power cut in 2014 and 2015. In addition, the tiered price led to more residents paying attention to and purchasing energy-saving appliances. A survey in Beijing in 2012 shows that after the implementation of the tiered price, about 60% of households choose to replace energy-saving appliances in a near future. Surveys from the appliance store also shows that more than 90% of customers asked the energy-saving attribute in the choice of home appliances, while before the introduction of the policy less than 30% of people would ask about it.

6. Conclusion

Against the backdrop of increasing emphasis on energy efficiency and energy conservation, this chapter reviews the background, characteristics, and effects of residential electricity pricing policies in China at different stages of development.

In the first stage (1987-1997), due to under-estimation of the role of electricity in economic development, increasing electricity demand with the improving living conditions, and the lagging behind of the electricity pricing reform, there was a lasting shortage of power supply which affected people's life and economic development tremendously.

In the second stage (1998-2003), because of the slowing down economy and the construction of the electricity industry for a decade, there was an unexpected surplus of power supply. The electricity pricing was therefore directed to encourage the electricity consumption from both households and industries.

In 2004, there was an unprecedented large-scale electricity shortage, which amounted to a gap up to 20 million kWh that is equivalent to the installed capacity of the Three Gorges Hydropower Station. To address challenges faced in the large-scale electricity shortage, the tiered electricity pricing reform was first carried out in three provinces, Zhejiang, Fujian and Sichuan, so as to evaluate the policy effects and obtain some experience before the nationwide implementation. While the policy effect of energy-saving was seen, the problems emerged from the pilot work were collected and discussed to prepare for a larger-scale of implementation.

However, due to various reasons, the nationwide implementation was stagnated during 2008-2010. Among all the reasons, "the 87 kWh red line" negated by the hearing and criticized widely was well-known. Nevertheless, the preparation for the future nationwide implementation continued in terms of efforts to installing the meters for every household.

Eventually, the tiered electricity pricing policy was implemented in 29 provinces in China since 1 July 2012. National Development and Reform

Commission set a general standard and the specific policy was made by the individual province based on its own economic and geographical conditions. Up to now, the tiered pricing policies are still undergoing some adjustments and changes.

REFERENCES

Cai, Qixing. & Shen, Bonian. (1987). On the relationship between tiered electricity pricing and the promotion of development of the electricity industry. *Energy Research and Information*, *3*(4), 7-12.

He, Xuemin. (2005a). The hearing to be held for tiered electricity pricing reform in Sichuan. *China Electric Power Newspaper*, on 25 November.

He, Xuemin. (2005b). Calling for hearing representatives and tiered electricity pricing expected out soon. *Southwestern Electric Power Newspaper*, on 4 December.

Liao, Xuezhong. & Bai, Jie. (2000). The report of the visit to Electric Bulb Company in Hong Kong. *Demand Side Management*, *2*, 44-46.

Luo, Xiangming. (2007). Do you use energy-saving lamps? – A survey on the usage of energy-saving bulbs in Chengdu Sichuan. *China Electric Power Newspaper*, on 29 March.

Lu, Shihong. (2008). Seminar on system building on data collection of electricity consumption of low-voltage users held by Stage Grid Corporation. *Demand Side Management*, *10*(6), 78.

Liang, Lu. (2013). *Policy evaluations of tiered electricity pricing in Beijing and the strategies*. Master thesis for North China Electric Power University.

Li, Dongmei. (2017). *Effect on residential electricity consumption of tiered pricing in Anhui and its energy-saving effect – An empirical study.* Master thesis for Zhejiang University of Finance and Economics.

Shi, Jiansuo. (2006). People applaud tiered electricity price. *China Electric Power Newspaper*, on 1 August.

Wei, He. (2009). Can tiered electricity price promote energy-saving? *China Daily*, on 27 November.

Wu, Chaoyang. (2011). Tiered electricity pricing in Zhejiang and its problems and solutions. *Energy Technology and Economics*, 5, 48-52.

In: Energy Policy
Editor: K. J. Sreekanth

ISBN: 978-1-53613-744-6

Chapter 4

BEHAVIOR CHANGE APPROACH: A METRIC TO PROMOTE AND SUSTAIN ENERGY EFFICIENCY

Fotouh Al-Ragom*[*], *PhD
Energy Efficiency Technologies Program,
Kuwait Institute for Scientific Research, Kuwait

ABSTRACT

The general trend in efficiency programs involves using technological and economical tools. These tools can support decision making specifically for program implementation justification (techno-economic validation). Once the programs are in place, the success of these programs lies in the proper adaptation and retained dedication of project stakeholders or end-users. Hence psychology and behavioral economics tools can be used to capture culture and psychology trends to devise comprehensive energy efficiency programs that can eliminate technologies adoption barriers. Such comprehensive programs will incorporate energy-focused behavior change interventions. This chapter reviews available energy-focused behavior change programs highlights

[*] Corresponding Author Email: fragom@kisr.edu.kw.

behavior biases and gaps and describes a behaviorally informed design scheme that can be utilized as a tool for developing energy efficiency behavior change interventions.

Keywords: behavior informed design, behavior change techniques, energy efficiency, design thinking

INTRODUCTION

"Changing behavior and reducing the per capita energy usage is tantamount to having found a new source of energy. Resource consumption and dependence offset in tandem rather than separately." (Frederiks, Stenner, Hobman, et al., 2015). Harnessing this new source can be of paramount effect to energy consumption and demand. Moreover, to ensure the realization of the maximum efficiency of a system, subsystems efficiencies must be optimized. In energy systems, subsystems will include system/technology adaption by end-users/consumers and their behavior change. Behavior change can propagate faster than infrastructure or technology transfer. It can also ensure the sustained use of systems and technologies. Behavior change is governed by behavioral economics and psychology. Understanding behavior change theory, norms, gaps and failure modes will provide insights into cognitive biases related to energy-related behavior hence explaining why individuals fail to commit to energy efficiency. Energy conservation behavior studies trace back to the 70s after the first Arab Oil Embargo (Shihata, 1974). These empirical studies show that to reform energy consumption behavior several factors need to be considered including rewards (financial) followed by feedback of behavioral performance, exhortation and information. The finical rewards are the most effective while information dissemination was found to be the least effective (Foxx1 & Hake, 1977), (Hayes & Cone, 1977). Kohlenberg et al. concluded that both cash incentives and feedback combined can maximize the reduction in electrical consumption while information dissemination such as the equipment power requirements had a very small affect on energy consumption patterns. They also noted that the removal of experimental treatments (ex. removal of energy consumption feedback device) reverted the consumption to pre- treatment patterns of consumption (Kohlenberg, Phillips, & Proctor, 1976).

In 1977, Milstein conducted an empirical study to assess the knowledge and behavior of the American consumers regarding energy conservation (Jeffrey S. Milstein, 1977). This study analyzed survey data gathered by the U.S. Federal Energy Administration in efforts to assess the effect of the Arab Oil embargo on energy conservation/consumption. The study spanned over 2 years (from 1974-1976) with 1000-1200 person/month were queried in 48 states through telephone calls (12-20 minutes). He evaluated several factors affecting behavior including the following:

- Awareness: A majority is aware of the U.S. energy problem due to the Oil embargo and that there is a public understanding of its implications yet, they will not take actions to conserve unless they "know how and why they should".
- Attitude: Milstein explained that what people say and what they do is totally different and that virtually everyone is for energy conservation but in the abstract. He elaborated that "there is a great discrepancy between lip service and action" and that people may not accurately report on their own behavior.
- Lack of Knowledge: He addressed the importance of lack of knowledge, as consumers sometimes do not know how to save energy and that there are some misconceptions.
- Other: He evaluated other factors including conveniences and comforts and concluded that Americans (while not a universal trend) highly value indulgence and comfortable life. This, in fact,, can apply to other nations around the world, in materialistic societies. He also evaluated scepticism and cynicism and showed that energy conservation actions are being mitigated because the consumers are a sceptic that their actions/sacrifices will be carried

equitably by others such as the government and the industry. He also highlighted the importance to use motives, the desire to financial benefits and to induce people to save energy (Jeffrey S. Milstein, 1977).

In recent studies such as the one by Frederiks et al. (Frederiks, Karen, & Hobman, 2015) the authors indicated that consumer choices and behavior are governed by cognitive biases and heuristics. They elaborated that people are driven by loss aversion rather than gain, use relative rather than absolute evaluation and prefer lower certainties over higher risks. According to the discussed studies and other several behavior studies, it has been shown that people can think and act differently and cannot maintain behavior change over a long period of time. Often a gap between people knowledge, value, attitude and intention and their observable energy consumption behavior is found. Hence people even when they know about the environmental benefits using a greener technology such as using solar panels, they do not adopt such a technology. Not adopting such a technology while knowing its benefits are not necessarily an indicator of low environmental values. This poses difficulties in predicting consumer behavior. While many think that household energy consumption is driven by financial incentives as indicated by Milstein (Jeffrey S. Milstein, 1977), other behavioral economics studies explain that other factors dictate consumers' behavior. These factors include biases such as the status quo, loss and risk aversion, sunk-cost effects, time and space (temporal and spatial) discounting, and the availability bias (Chatzidakis & Lee, 2013). Thus the behavior assessment is a vast field. Many factors can contribute to either promoting or hindering a targeted behavior. This chapter will shed some light on some factors that can be of influence on behavior in anticipation that it can provide a better understanding of behavior physiological factors. This understanding can aid in designing adaptable energy efficiency interventions that can aid in developing a targeted long-lasting behavior change.

BEHAVIOR GAPS

Despite the individuals' awareness, values, preferences and intention towards energy efficiency, sustainability or any socially desirable behavior, their actions do not coincide with their intentions and their values and attitudes fail to materialize into technologies purchases. Connolly and Prothero's argued that while individuals feel empowered and responsible for environmental issues, their insecurities of not knowing what the 'best' or 'right' choices are can hinder their involvement in pursuing or adopting options that are inline with best environmental practice (Connolly & Prothero, 2008). Often attitudes, intentions and behavior do not correlate well, which can lead to a low adoption rate of energy-saving measures by the consumers (Baker, 2003). This can hamper the desired effect from awareness campaigns, which assume that increasing the consumers' knowledge will alter their behavior and guarantee long-term adoption of efficient practices. Hence the analysis of the behavior gaps can be used to understand the energy consumption behavior and consumer preferences. Various behavior gaps exist, which can aid, once understood in designing successful behavior change campaigns as shown in Figure 1.

Attitude-Action Gap

This gap represents the discrepancy between consumers' expressed preferences for energy efficiency practice or technology adoption and their actual willingness/unwillingness to commit is commonly referred to as the attitude-behavior/action gap (Barnett, 2007; Clarke, Barnett, Cloke, & Malpass, 2007; Claudy, Peterson, & Driscoll, 2013; Moraes, Carrigan, & Szmigin, 2012). The attitude-behavior gap was addressed by several authors and many have demonstrated that attitude-behavior inconsistencies exist as some case studies have shown that positive attitudes towards behaviors will not necessarily lead to the performance of the intended behavior. Marius et al. argued that when focusing only on attitude-behaviour, reasons against adoption will not be evident (Claudy et al.,

2013). By applying behavioral reasoning theory on consumer adoption of solar panels they have concluded that it is important to reduce reasons against adoption in order to increase the adoption of the panels and other renewable energy technologies in the future. Chatzidakis and Lee supported the phenomena of negation and affirmation in reasons theory as they highlighted how the "reasons against" consumption are not always the logical opposite of the "reasons for" consumption (Chatzidakis & Lee, 2013).

They have explained that the reasons against adoption of solar panels included the high cost of renewable energy technologies due to the conventional energy government direct/indirect subsidies, and discussions of stakeholders on the financial payback time of renewable energy investment decisions (Claudy et al., 2013).

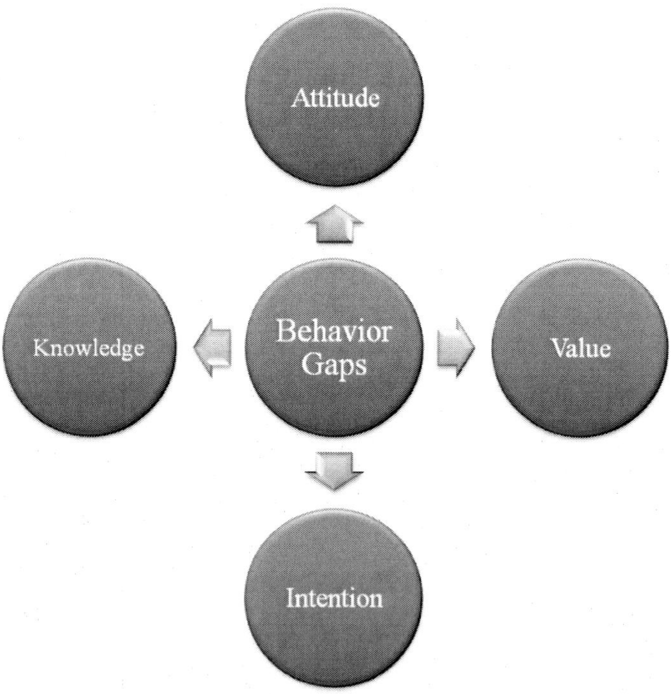

Figure 1. Identified behavior gaps.

Knowledge-Action Gap

Verplanken and Wood argued that in the long run informational interventions will not succeed. They stressed the importance of sustaining habits for behavior change by considering the socio-structural factors. They suggest that repeatability can lead to decision withdrawal and that behavior can be mechanically prompted by contextual and environmental factors (Verplanken & Wood, 2006). Newton and Meyer pointed that "attitudes and behaviors are highly dynamic which prompts changes at any point in time" (Newton & Meyer, 2013). Psychological barriers, not knowing how to change, is amplified in high-cost energy conservation behavior. Lacroix and Gifford (Lacroix & Gifford, 2017) explain that ignorance (as a barrier to behavior change) might not be a barrier for a low-cost behavior (ex. taking shorter showers) but it may apply in a high-cost behavior (ex. installing solar panels).

Intention-Action Gap

Newton and Meyer, 2013, illustrated that, in the conceptual framework for consumption research, urban locality, dwelling, household and individuals all contribute to the actual consumption. While the determinants of consumption include the demand side (individuals, households and the settings related to their dwelling, urban location or social surroundings) and the supply side (a technological innovation of key infrastructures, systems and services), the individuals and households behavior change can hold a faster rate of sustainability transformation. They implied that the change is possible in both determinants, yet the timescale required is considerable for the supply side which explains the amplified government's interest in behavior change policies and programs (Newton & Meyer, 2013; UNEP, 2011). In addition, they highlighted that several research studies on urban resource use are finding a gap between attitudes and actions as individuals' behavior and their intentions (Gilg & Barr, 2005; Lorenzoni, Nicholson-Cole, & Whitmarshb, 2007).

Newton and Meyer used the lifestyle segmentation (LSS) in studying actual consumption behavior. While they have detected three clusters "committed greens", "material greens" and enviro-sceptics" which validated typology, they could not detect any consumption behavior variation across the three clusters. They reasoned that the current gap between intentions and action is due to the range of barriers to sustainable consumption. These barriers include lack of relevant knowledge; availability of information; organizational challenges, i.e., how to get it done; time (temporal) constraints, similar to (Chatzidakis & Lee, 2013); and financial constraints (return on investment).

Value-Action Gap

This gap describes how individuals place a high value on an issue and the low level of action they take to counteract in an attempt to reform the issue in concern. Environmental behavior studies demonstrate that, in view of climate change, even though Americans and Canadians citizens think that climate change is a serious problem and that scientific evidence is proving the link between human activity and the climate change, they are still engaged in activities that lead to GHG emissions (Lacroix & Gifford, 2017; Leiserowitz, 2006; Lorenzoni et al., 2007). Blake theorized that, based on research evaluation, " that people do not have a fixed, rational and ready-made set of values that will be activated by particular calls to action; rather people's values are negotiated, transitory and sometimes contradictory" (Blake, 1999). From baseline research based on questionnaires and face-to-face interviews in Huntingdonshire sustainable community project in which respondents despite their general concern for the environment, were asked to identify the barriers or reasons that prevented them from carrying out particular environmental actions. The responses were summarized and coded. Three different categories of obstacles were identified between concern and action. These categories are as follows:

- Individuality: Personal attitudes or cognitive structure can prevent an individual from taking action because of several reasons including laziness, lack of interest, not having a self-belief in being the right person for action.
- Responsibility: The belief that external (government, groups) control rather than internal control (individual) can create a difference in changing the situation is another barrier to action. The lack of responsibility is also amplified in those individuals who do not own the property they live in. They lack the motivation to take action even if they value environmental concerns related to rational consumption.
- Practicality: Some practical constraints (social or institutional) may prevent people from adopting an action, irrespective of their values or intentions. These include lack of time, money, physical space, information, encouragement and experience (UK Essays 2015).

Thus, according to Blake, the importance of different barriers to action will vary widely in different communities. Different appropriate strategies need to be designed for specific individuals, communities and institutions. As he clearly states that "There can never be a blueprint for encouraging environmental action".

The behavior gaps discussed above can hinder any behavior intervention program. Hence understanding these gaps and mitigating their effects is essential to interventions success. These gaps usually are happening on the subconscious level. Thus if such gaps are pointed out to the targeted individuals, they might be more aware of their actions which might lead to a behavior change that can match their values and attitudes.

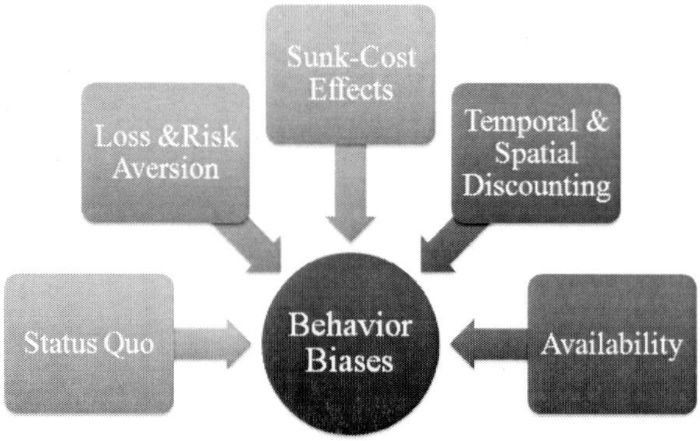

Figure 2. The most pervasive behavior biases.

BIASES

A stimulus can alter an intended behavior of humans by invoking certain biases. Accordingly, understanding these biases can help in adjusting behavior to achieve the desired target such as promoting sustainable energy and energy efficiency. Figure 2 illustrates the most powerful human biases including status Quo, loss and risk aversion, sunk cost effects, temporal and spatial discounting, and availability. While other biases exist the most pervasive one will be discussed. These are important as a large body of research indicates that biases can affect individuals' decisions even when cost-benefit shows advantageous options.

Status Quo

Humans tend to go with the flow and resist change (i.e., Status Quo). Individuals prefer to follow the status quo even when new alternatives might be of financial benefits or materially advantageous (Frederiks, Stenner, Hobman, et al., 2015; Kahneman, Knetsch, Thaler, Johnson, & Professor, 1991; Kahneman & Tversky, 1979; Samuelson & Zeckhauser, 1988). When individuals get overwhelmed with too many options, they usually prefer the easier option which is, choose nothing rather than face the risk of the new and unknown. "People don't like change, and they often don't choose to change a current choice, even if the options available are better than the status quo" (Jewel 2015). Understanding this bias can aid in designing well-suited programs for energy efficiency through behavior change. Opower is an example of a company that utilized behavioral economics in the energy field. Opower, owned by Oracle, is a provider of customer engagement and energy efficiency cloud services to utilities (Oracle, 2016). Opower provides customer engagement platforms for utilities. Opower products are being used by more than 100 utilities in 9 countries. More than 50 million homes are being served by Opower through demand-side management platforms. Opower management platforms which were developed using design thinking methodology informs customers about their energy consumption, and provide personalized ways to save energy and money. Thus their focus is on tailoring solutions to serve the needs of their customers.

Loss and Risk Aversion

The loss/risk aversion in behavior economics arise from prospect theory. Prospect theory is a behavioral model defined by Kahneman and Tversky in 1979 (Kahneman & Tversky, 1979). It explains how people choose between alternatives under risk and uncertainty. They theorized that people value gains and losses differently, and, will base decisions on perceived gains rather than perceived losses. People value certain outcomes rather than taking risks for higher gains involving some uncertainty, this is termed as the certainty effect. The certainty effect contributes to lose/risk aversion. Risk aversion is in selecting choices that involve sure gains and loss aversion (avoid taking risks) in choices that involve sure losses. Thus, if a person is given two equal choices, one expressed in terms of possible gains and the other in possible losses, he/she would choose the former, even when they achieve the same economic end result.

Sunk Cost Effect

Once a valued resource such as effort, time, and money is invested, people tend to become fixated (persistent) in an endeavour and avoid new opportunities even if the valued recourse becomes risky. This is known as the "sunk cost" (Arkes & Blumer, 1985; Garland, 1990; Kahneman et al., 1991; Newport, 1991; Thaler, 1980). Individuals would prefer to recover losses already suffered and discount future costs and benefits. An example of that is when keeping a losing stock and holding it hoping that it will recover. Fredericks et al. (Frederiks, Stenner, Hobman, et al., 2015) argued that this might stem from "Don't waist rule" learned through childhood, where abandoning an option is like waiting (ex. food). Based on research findings it has been suggested that the rate of use of a good or a service increases as the purchase value increases (Arkes & Blumer, 1985). As an example enduring costs for purchasing cars, appliances can increase their use. Hence cars are driven more; appliances are used more often even if not needed (Wolfram 2013). Dixit and Pindyck (1994) proposed a model of investment behavior. This model explains that procrastination of technology adoption when faced with future development uncertainty, can be a rational option. Van Soest and De Groot empirically tested and validated that model (van Soest and de Groot 2011).

Temporal and Spatial Discounting

Discounting can be defined as the "technique for comparing the values of costs and benefits which occur at different points in time" (Baum, Easterling, Baum, & Easterling, 2010). Temporal (time) or Spatial (space, distance), discounting include comparing values of an opportunity when considering timeframe or distance (location, space availability). Temporal and spatial discounting are sometimes linked and both can affect the perception of consequences of actions (Baum, 2012; Peattie, 2010). Analyzing temporal and special discounting can help in understanding how humans discount across space and time. Decisions are affected when considering the timeframe of any given choice or option. Temporal frame whether it's daily, monthly or yearly contributes to loss aversion. Energy efficiency measures require initial investments, while the benefits are realized later in time. Hence when framing any customer-focused messages, temporal discounting need to be taken into consideration, to minimize the effect of the initial cost incurred and highlight the future savings that will be realized.

Gill found that framing energy savings opportunity on a monthly basis can encourage its adaptation (Gill, 2015) and reduce the temporal discounting. The "pennies-a-day" strategy, in which an aggregate expense is represented on small daily or ongoing expense can provide and explanation of Gill's finding (Gourville, 1998). This concept explains why customers prefer monthly subscription of magazines, internet or mobile plans rather than the yearly subscription. Delayed gratification can also be offset by either providing small immediate rewards (ex. praise, recognition, in-kind gifts) or using a technique known as 'episodic future thinking' (Frederiks, Stenner, & Hobman, 2015). Episodic future thinking involves describing future rewards in a more vivid and tangible way (Hershfield et al., 2011; Kim, Schnall, & White, 2013; Oluyomi Daniel, Stanton, & Epstein, 2013).

Socio-spatial distance is the effect of the resistance on the person willingness to take action. Busse and Menzel examined the impact of social-spatial distance adolescents' willingness to engage in pro-environmental behavior (Busse & Menzel, 2014).

The literature on spatial discounting related to various issues can highlight the link between spatial discounting and risk perception. Baum et al. showed that climate change adaptation is affected by spatial discounting (Baum et al., 2010). People spatially discount the instrumental value of different locations which led some to migrate to these locations. Horst evaluated spatial discounting on wind farm acceptance. He explained that while the concept of some services (i.e., renewable energy) is accepted by the studied population in principle, the residents opposed such service in practice. This phenomenon is known as; not in my back yard (NIMBY) behavior. He elaborated that NIMBY was affected by the living proximity of the population (Van Der Horst, 2007). Hence those living close to the site of a proposed windfarm are opposed to it because of the level of risk perception. On the other hand, those who lived further away from an existing windfarm are also opposed to it as they lack the local experience which can affect their perception benefits. He added that once the projects were started the opposition rate dropped. Thus, spatial discounting, as well as temporal discounting, is not static but dynamic in nature.

Design Thinking Methodology

> "We cannot solve problems by using the same kind of thinking we use when we created them".
>
> *Albert Einstein*

Commonly a system, device, technology or intervention is developed to serve a certain need or solve an existing problem then once developed; it has to be marketed to push towards its utilization. On the other hand, People, the end users, are the core of design thinking. People are the users of buildings, equipment and new gadgets. Hence, design thinking is an approach that puts "human needs, capabilities and behavior in the first place, and then applies technical and design tools to accommodate those needs and ways of behaving." (Kuzniatsou 2016). Design thinking invokes innovation and creativity. It is defined by Tim Brown CEO of IDEO, an international design and consulting firm in California USA, as follows:

"Design thinking is a human-centred approach to innovation that draws from the designer's toolkit to integrate the needs of people, the possibilities of technology, and the requirements for business success." (IDEO 2017).

Services, products, devices developed using design thinking, will resonate more deeply with people making them easily adaptive for a longer period. The origin of design thinking can be linked to "The Sciences of the Artificial," Nobel Prize laureate Herbert Simon, published in 1969 (Herbert A. Simon, 1996). He developed the first formal 7 steps framework for design thinking. Currently, there are different alternatives to the number of steps ranging from 3 to 7. While there is no standardized model for design thinking, each industry develops it's own using a basic framework that is common to all design thinking processes. The practices that were developed at d.school at Stanford University involve a 5 steps model will be discussed here (Stanford, 2017). This framework consists of the following five steps (Rikke and Teo, 2017-a):

- Empathize: In simple terms, empathy means the ability to feel and understand the feelings of others. Thus emphasize mode is what a person can do to understand people needs (physical and emotional). Since usually when a system developer or designer work to solve a problem it is rarely his/her own. Accordingly, the empathy must be built for the users in order to understand how they feel and think. One can develop empathy through observing others behavior, engaging with users through questionnaires and interviews and immersing in the user's experience by actually living in the situation. By doing that, insights can be revealed that can lead to innovative solutions.
- Define: After developing empathy for the user, and based on the gathered information, this step is all about making sense and defining the problem statement and then coming up with an actionable problem statement. This stage will start the process for the following stage "Ideation" by asking the questions "How might we do ….?" (SAP 2017).
- Ideate: This step is all about generating a wide range of ideas. It is not about coming up with the best one, as this comes at a later stage. Thus the focus is ideas generation to create, think outside the box, (innovative) solutions for the users. There are numerous techniques for ideation stage including Brainstorming, bodystorming, stoking, worst possible idea and SCAMPER (Substitute, Combine, Adapt, Modify/Magnify/Minify, Put to another use, eliminate and Reverse) (Rikke and Teo 2017-b). The working team can vote on three ideas to carry them out for the next step that is prototyping.

- Prototype: In this step, ideas are transferred to the physical world. The physical forms can be post-it-notes, role playing, an object, a storyboard or an interface. The prototypes allow the users and the designers to interact with the idea in design to test functionality. This is an experimental stage to help in selecting the best solution. The solutions can be rejected, redesigned or accepted based on the interaction and user experience. Rejected/redesigned solutions can help in reformulating the problem statement to come up with a better solution if none has been achieved. At this stage, the design team will be well informed about the product problems, how the users will feel, think and interact with the end product.
- Test: This mode will allow receiving feedback on the created prototypes from the end users. If the users like the prototype, then you need to know why. The prototype is tested by end-users and the design team can monitor how it is being used or misused, how they interact with it and listen to what they say about it and note the questions they have about it. Testing allows refinement of the solutions to make them better.

An example of an energy saving product that was developed using design thinking is the NEST thermostat. The thermostat main design driver was the frustration of programming old thermostats. This driver guided Tony Fadell, a former Senior Vice President at Apple, who was building an energy efficient house at that time. Tony engaged experts, contractors, conducted research, evaluated other versions of thermostats, and interviewed users to come up with the NEST. NEST through its learning algorithm known as "Autoschedule" develops a customer-tailored schedule. The development of NEST is an example of design thinking approach that is being used in energy efficiency (Nest Labs 2017). D. light is another example of employing design thinking in energy efficiency. D. light is a company with a mission to: "enable households without reliable electricity to attain the same quality of life as those with electricity." The founders of the company developed their idea during a design thinking course at Stanford University. They have arrived at their design while attending a design thinking class at Stanford University titled Design for Extreme Affordability. They immersed themselves in lives of rural villagers. They have realized the dependence of rural villagers on Kerosene lamps despite their cost, indoor air pollution and contribution to causing a fire. Their solution was to bring clean, safe and bright light to every rural household. Their products which include a range of modular products such as desk lights, portable lanterns and solar home systems were sold in about 60 countries and improving the lives of nearly 30 million people (D.light design. 2018).

From the examples discussed above, it can be noticed that design thinking can play a vital role in the development of programs, products and systems. It can increase the adaptability of products. In the case of NEST thermostat, through surveying their customers they found out that "99% of those devices have schedules that save energy, as opposed to 11% before" (Busari 2012).

The Ripple Effect Concept

The "ripple effects" is concept studied by researchers in the health field. This concept addresses the behavioral interventions and its effect on triggering multiple outcomes that are beyond the main target of the behavior interventions (Wilson, 2015). These outcomes, in the behavioral medicine field, include physical, psychological, and social health domains across the lifespan of individuals. Numerous evidenced-based intervention trials supported the existence of ripple effects. Some of these trials involved interventions in obesity for children and adolescence, drug prevention, diabetes and self-regulation of chronic diseases (ex. blood pressure) (Davis et al., 2011; Eather, Morgan, & Lubans, 2013; Epstein et al., 2008; Hillman et al., 2009; van Stralen et al., 2012; Wilson et al., 2009; Zarrett, Skiles, Wilson, & McClintock, 2012). The findings of these studies Showed evidence of ripple effect such as significant improvements in memory, attention, executive functioning (ex. planning and decision-making) arising from more complex cognitive processing by using measures of neuroelectric changes during cognitive performance tasks and self-regulation processes due to long-term adherence to healthy lifestyle behaviors engagement (Ahn et al., 2013; Botvin, Gilbert J, Griffin, Kenneth W, Paul, Elizabeth, Macaulay, Araxi, 2003; Botvin & Griffin, 2004; Botvin, Griffin, & Nichols, 2006; Gerber, 2006; Gorin et al., 2009; Kerr et al., 2008; Mata et al., 2009; Powers, Olsen, Oddone, & Bosworth, 2009; Werch et al., 2007). In addition, positive decision-making enhanced social and personal competence skills. These cascading effects across physical, mental, and social outcomes of well-being can lead to a higher return on investment (ex. less healthcare expenditure).

In view of energy efficiency can realize multiple benefits (i.e., ripple effects) include increased productivity enhanced product quality, improved system reliability and environmental compliance (Russell, 2015). Enhanced learning is an example of such effects (Howard, 2016). Usually, energy efficiency improvements focus almost exclusively on energy savings. While other benefits are recognized in concept but not highly portrayed. In Norway, a school installed tunable LED lighting. These are considered human-centric lighting (HCL) that are known to promote health and wellbeing (Wright, 2017). Lighting replacement additional benefits can include (if lights with longer life are used CFL vs. LED) fewer replacements and lower maintenance costs (Sullivan, Armel, & Todd, 2012). In the manufacturing process, monitoring energy can lead to conservation of water and raw materials and reduced waste (ACEEE, 2015).

CONCLUSION

Energy consumption behavior change targeting achieving energy efficiency is a field that spans over numerous fields including energy economics, behavior change technology, behavior physiology and theories. To create intervention programs that can be the adapted life style in the future requires tremendous efforts specifically from the scientific community. It has been shown throughout this chapter how this filed can benefit from other fields such as psychology, marketing, health and education. Efforts are required to reduce the perception of psychological barriers to change individual and societal behavior. Combinations of voluntary, regulatory, and structural approaches must be incorporated in interventions to ensure their adaptation and reap their fruition. More research can be targeted to understand behavior science as in other fields to device guidelines that can be developed to direct energy behavior towards the favoured paths.

Several studies are available that benefit from behavior science and theories in sustainability and environment perspective. Yet more specific ones targeting energy conservation/efficiency are required. The understanding of biases, behavior gaps, ripple effects and design thinking concepts can feed into designing consumer (human) centered approaches to come up with a comprehensive behaviorally informed design schemes for developing energy systems, technologies, awareness campaign and behavior interventions that can be adapted for a longer period of time or even become a lifestyle.

REFERENCES

ACEEE. (2015). *How energy efficiency investment creates a ripple effect of multiple benefits for businesses beyond energy savings* | ACEEE. Retrieved December 24, 2017, from https://aceee.org/blog/2015/01/how-energy-efficiency-investment-crea.

Ahn, S., Basu, R., Smith, M. L., Jiang, L., Lorig, K., Whitelaw, N., & Ory, M. G. (2013). The impact of chronic disease self-management programs: healthcare savings through a community-based intervention. *BMC Public Health*, *13*(1), 1141. http://doi.org/10.1186/1471-2458-13-1141.

Arkes, H. R., & Blumer, C. (1985). The Psychology of Sunk Cost. *Organizational behavior and human decision processes*, *35*, 124–140. Retrieved from http://citeseerx.ist.psu.edu/viewdoc/download?doi=10.1.1.452.2318&rep=rep1&type=pdf.

Baker, M. J. (2003). *The Marketing Book* (5th ed.). Linacre House. Retrieved from http://htbiblio.yolasite.com/resources/Marketing Book.pdf.

Barnett, C. (2007). *Governing the Subjects and Spaces of Ethical Consumption*. Swindon, UK. Retrieved from http://cftn.ca/sites/default/files/AcademicLiterature/Governing the Subjects and Spaces of Ethical Consumption.pdf.

Baum, S. D. (2012). *Discounting across space and time in climate change assessment.* A Dissertation in Geography. The Pennsylvania State University. Retrieved from http://sethbaum.com/ac/2012_ Dissertation.pdf.

Baum, S. D., Easterling, W. E., Baum, S. D., & Easterling, W. E. (2010). Space-time discounting in climate change adaptation. *Mitig Adapt Strateg Glob Change*, *15*, 591–609. http://doi.org/10.1007/s11027-010-9239-9.

Blake, J. (1999). Local Environment Overcoming the "value-action gap" in environmental policy: Tensions between national policy and local experience Overcoming the "Value-Action Gap" in Environmental Policy: tensions between national policy and local experience. *Local Environment*, *4*(3), 257–278. http://doi.org/10.1080/13549839908 725599.

Botvin, Gilbert J, Griffin, Kenneth W, Paul, Elizabeth, Macaulay, Araxi, P. (2003). Preventing Tobacco and Alcohol Use Among Elementary School Students Through Life Skills Training. *SubStance*, *652*(March 2015), 37–41. http://doi.org/10.1300/J029v12n04.

Botvin, G. J., & Griffin, K. W. (2004). Life Skills Training: Empirical Findings and Future Directions. *The Journal of Primary Prevention*, *25*(2). Retrieved from http://citeseerx.ist.psu.edu/viewdoc/download?doi=10.1.1.552.5881&rep=rep1&type=pdf.

Botvin, G. J., Griffin, K. W., & Nichols, T. D. (2006). Preventing youth violence and delinquency through a universal school-based prevention approach. *Prevention Science*, *7*(4), 403–408. http://doi.org/10.1007/s11121-006-0057-y.

Busse, M., & Menzel, S. (2014). The role of perceived socio-spatial distance in adolescents' willingness to engage in pro-environmental behavior. *Journal of Environmental Psychology*, *40*, 412–420. http://doi.org/10.1016/j.jenvp.2014.10.002.

Chatzidakis, A., & Lee, M. S. W. (2013). Anti-Consumption as the Study of Reasons against. *Journal of Macromarketing*, *33*(3), 190–203. http://doi.org/10.1177/0276146712462892.

Clarke, N., Barnett, C., Cloke, P., & Malpass, A. (2007). The Political Rationalities of Fair-Trade Consumption in the United Kingdom. *Politics & Society*, *35*(4), 583–607. http://doi.org/10.1177/0032329207308178.

Claudy, M. C., Peterson, M., & Driscoll, A. O. (2013). *Understanding the Attitude-Behavior Gap for Renewable Energy Systems Using Behavioral Reasoning Theory*, *33*(4), 273–287. http://doi.org/10.1177/0276146713481605.

Connolly, J., & Prothero, A. (2008). Green Consumption Life-politics, risk and contradictions. *Journal of Consumer Culture*, *8*(1), 117–145. http://doi.org/10.1177/1469540507086422.

D.light design. 2018. http://www.dlight.com/.

Davis, C. L., Tomporowski, P. D., Mcdowell, J. E., Austin, B. P., Miller, P. H., Allison, J. D., … Naglieri, J. A. (2011). Exercises improves executive function and achievement and alters brain activation in overweight children A randomized controlled trial. *Nih*, *30*(1), 91–98. http://doi.org/10.1037/a0021766.Exercise.

Eather, N., Morgan, P. J., & Lubans, D. R. (2013). Social support from teachers mediates physical activity behavior change in children participating in the Fit-4-Fun intervention. *International Journal of Behavioral Nutrition and Physical Activity*, *10*(1), 68. http://doi.org/10.1186/1479-5868-10-68.

Epstein, L. H., Roemmich, J. N., Robinson, J. L., Paluch, R. A., Winiewicz, D. D., Fuerch, J. H., & Robinson, T. N. (2008). A Randomized Trial of the Effects of Reducing Television Viewing and Computer Use on Body Mass Index in Young Children. *Arch Pediatr*

Adolesc Med, *162*(3), 239–245. Retrieved from https://www.ncbi.nlm.nih.gov/pmc/articles/PMC2291289/pdf/nihms43558.pdf.

Foxx1, R. M., & Hake, D. F. (1977). Gasoline conservation: a procedure for measuring and reducing the driving of college students. *Journal of applied behavior analysis*, *10*(1), 61–74. Retrieved from https://www-ncbi-nlm-nih-gov.colorado.idm.oclc.org/pmc/articles/PMC1311150/pdf/jaba00112-0063.pdf.

Frederiks, E. R., Karen, S., & Hobman, E. V. (2015). Household energy use: Applying behavioural economics to understand consumer decision-making and behaviour. *Renewable and Sustainable Energy Reviews*, *41*, 1385–1394. Retrieved from https://ac.els-cdn.com/S1364032114007990/1-s2.0-S1364032114007990-main.pdf?_tid=5c4f815a-e275-11e7-8cfd-00000aab0f01&acdnat=1513438114_9df604de6660dc327c99b5b5a970cc81.

Frederiks, E. R., Stenner, K., Hobman, E., Frederiks, E. R., Stenner, K., & Hobman, E. V. (2015). *Household energy use : Applying behavioural economics to understand consumer decision- making and behaviour consumer decision-making and behaviour,* (January). http://doi.org/10.1016/j.rser.2014.09.026.

Frederiks, E. R., Stenner, K., & Hobman, E. V. (2015). Household energy use: Applying behavioural economics to understand consumer decision-making and behaviour. *Renewable and Sustainable Energy Reviews*, *41*, 1385–1394. http://doi.org/10.1016/j.rser.2014.09.026.

Garland, H. (1990). Throwing Good Money After Bad: The Effect of Sunk Costs on the Decision to Escalate Commitment to an Ongoing Project. *Journal of Applied Psychology*, *75*(6), 728–731. Retrieved from http://www.communicationcache.com/uploads/1/0/8/8/10887248/throwing_good_money_after_bad-_the_effect_of_sunk_costs_on_the_decision_to_esculate_commitment_to_an_ongoing_project.pdf.

Gerber, B. S. (2006). The chronic disease self-management program: extending reach through the internet. *Medical Care*, *44*(11), 961–3. http://doi.org/10.1097/01.mlr.0000244923.80923.1e.

Gilg, A., & Barr, S. (2005). Behavioural attitudes towards water saving? Evidence from a study of environmental actions. *Ecological*

Economics, *57*, 400–414. http://doi.org/10.1016/j.ecolecon. 2005.04. 010.

Gill, C. (2015). Temporal Framing, Endowment, and Energy Choices. Retrieved December 16, 2017, from http://www.sjdm.org/ presentations/2015-Poster-Gill-Carrie-LossAversion-Framing-Energy. pdf.

Gorin, A. A., Wing, R. R., Fava, J. L., Hospital, T. M., Street, R., Jakicic, J. M., ... Brelje, K. (2009). *Environment : Evidence of a ripple effect, 32*(11), 1678–1684. http://doi.org/10.1038/ijo.2008.150.Weight.

Gourville, J. T. (1998). Pennies-a-Day: The Effect of Temporal Reframing on Transaction Evaluation. *Journal of Consumer Research*, *24*(4), 395–403. http://doi.org/10.1086/209517.

Hayes, S. C., & Cone, J. D. (1977). Reducing residential electrical energy use payments, information, and feedback. *Journal of Applied Behavior Analysis*, *10*(3), 425–435. Retrieved from https://www-ncbi-nlm-nih-gov.colorado.idm.oclc.org/pmc/articles/PMC1311206/pdf/jaba00114-0053.pdf.

Herbert A. Simon. (1996). *The Sciences of the Artificial*. MIT Press. Retrieved from https://monoskop.org/images/9/9c/Simon_Herbert_A_The_Sciences_of_the_Artificial_3rd_ed.pdf.

Hershfield, H. E., Goldstein, D. G., Sharpe, W. F., Fox, J., Yeykelis, L., & Bailenson, J. N. (2011). Increasing saving behavior through age-progressed renderings of the future self. *Journal of Marketing Research*, *48*, S23–S37. http://doi.org/10.1509/jmkr.48.SPL.S23.

Hillman, C. H., Pontifex, M. B., Raine, L. B., Castelli, D. M., Hall, E. E., & Kramer, A. F. (2009). The effect of acute treadmill walking on cognitive control and academic achievement in preadolescent children. *Neuroscience*, *159*, 1044 –1054. http://doi.org/10.1016/j.neuroscience. 2009.01.057.

Howard, J. (2016). *How Just The Right Lighting May Improve Learning In Classrooms* | HuffPost. Retrieved November 1, 2017, from https://www.huffingtonpost.com/entry/lighting-boost-learning-concentration_us_5720cb14e4b0b49df6a9b73e.

Jeffrey S. Milstein. (1977). Attitudes, Knowledge and Behavior of American Consumers Regarding 1977. *Advances in Consumer Research*, *4*, 315–321. Retrieved from http://acrwebsite.org/volumes/9358/volumes/v04/NA-04.

Kahneman, D., Knetsch, J. L., Thaler, R. H., Johnson, H., & Professor, L. (1991). Anomalies: The Endowment Effect, Loss Aversion, and Status Quo Bias Anomalies The Endowment Effect, Loss Aversion, and Status Quo Bias. *The Journal of Economic Perspectives Journal of Economic Perspectives*, *5*(1), 193–206. Retrieved from https://www.princeton.edu/~kahneman/docs/Publications/Anomalies_DK_JLK_RHT_1991.pdf.

Kahneman, D., & Tversky, A. (1979). Prospect theory: An analysis of decision under risk. *Econometrica*, *47*(2), 263–291. Retrieved from https://www.princeton.edu/~kahneman/docs/Publications/prospect_theory.pdf.

Kerr, J., Patrick, K., Norman, G., Stein, M. B., Calfas, K., Zabinski, M., & Robinson, A. (2008). Randomized control trial of a behavioral intervention for overweight women: Impact on depressive symptoms. *Depression and Anxiety*, *25*(7), 555–558. http://doi.org/10.1002/da.20320.

Kim, H., Schnall, S., & White, M. P. (2013). Similar Psychological Distance Reduces Temporal Discounting. *Personality and Social Psychology Bulletin*, *39*(8), 1005–1016. http://doi.org/10.1177/0146167213488214.

Kohlenberg, R., Phillips, T., & Proctor, W. (1976). A behavioral analysis of peaking in residential electrical-energy consumers. *Journal of Applied Behavior Analysis*, *9*(1), 13–18. Retrieved from https://www-ncbi-nlm-nih-gov.colorado.idm.oclc.org/pmc/articles/PMC1311891/pdf/jaba00051-0015.pdf.

Lacroix, K., & Gifford, R. (2017). Psychological Barriers to Energy Conservation Behavior : The Role of Worldviews and Climate Change Risk Perception. *Environment and Behavior*, 1–32. http://doi.org/10.1177/0013916517715296.

Leiserowitz, A. (2006). Climate change risk perception and policy preferences: The role of affect, imagery, and values. *Climatic Change*, *77*(1–2), 45–72. http://doi.org/10.1007/s10584-006-9059-9.

Lorenzoni, I., Nicholson-Cole, S., & Whitmarshb, L. (2007). Barriers perceived to engaging with climate change among the UK public and their policy implications. *Global Environmental Change*, *17*, 445–459. Retrieved from https://ac-els-cdn-com.colorado.idm.oclc.org/S0959378007000209/1-s2.0-S0959378007000209-main.pdf?_tid=d639c4ba-d476-11e7-b0d5-00000aab0f6b&acdnat=1511899432_2d3898b5a4dbba9b766af5d555f895d8.

Mata, J., Silva, M. N., Vieira, P. N., Carraça, E. V., Andrade, A. M., Coutinho, S. R., … Teixeira, P. J. (2009). Motivational "spill-over" during weight control: Increased self-determination and exercise intrinsic motivation predict eating self-regulation. *Health Psychology*, *28*(6), 709–716. http://doi.org/10.1037/a0016764.

Moraes, C., Carrigan, M., & Szmigin, I. (2012). The coherence of inconsistencies: Attitude-behaviour gaps and new consumption communities. *Journal of Marketing Management*, *28*(1–2), 103–128. http://doi.org/10.1080/0267257X.2011.615482.

Newport, S. (1991). Effects of Absolute and Relative Sunk Costs on the Decision to Persist with a Course of Action. *Organizational Behavior and Human Decision Processes*, *48*, 55–69. Retrieved from https://www.gwern.net/docs/sunkcosts/1991-garland.pdf.

Newton, P., & Meyer, D. (2013). Exploring the attitudes-action gap in household resource consumption: Does "Environmental Lifestyle" segmentation align with consumer behaviour? *Sustainability (Switzerland)*, *5*(3), 1211–1233. http://doi.org/10.3390/su5031211.

Oluyomi Daniel, T., Stanton, C. M., & Epstein, L. H. (2013). The Future Is Now: Reducing Impulsivity and Energy Intake Using Episodic Future Thinking. *Psychol Sci.*, *24*(11), 2339–2342. http://doi.org/10.1177/0956797613488780.

Peattie, K. (2010). Green Consumption: Behavior and Norms. *Annual Review of Environment and Resources*, *35*(1), 195–228. http://doi.org/10.1146/annurev-environ-032609-094328.

Powers, B. J., Olsen, M. K., Oddone, E. Z., & Bosworth, H. B. (2009). The Effect of a Hypertension Self-Management Intervention on Diabetes and Cholesterol Control. *American Journal of Medicine*, *122*(7), 639–646. http://doi.org/10.1016/j.amjmed.2008.12.022.

Russell, C. (2015). *Multiple Benefits of Business-Sector Energy Efficiency : A Survey of Existing and Potential Measures*. Washington, DC..

Samuelson, W., & Zeckhauser, R. (1988). Status Quo Bias in Decision Making. *Journal of Risk and Uncertainty*, *1*, 7–59. Retrieved from https://link-springer-com.colorado.idm.oclc.org/content/pdf/10.1007%2FBF00055564.pdf.

Shihata, I. F. I. (1974). Destination embargo of arab oil its legality under international law. *American Journal of International Law*, *68*(4), American Journal of International Law. http://doi.org/10.3366/ajicl.2011.0005.

Stanford (2017). Hasso Plattner Institute of Design at Stanford University. https://dschool.stanford.edu/about/

Sullivan, D., Armel, C., & Todd, A. (2012). *When "Not Losing" is Better Than "Winning:" Using Behavioral Science to Drive Customer Investment in Energy Efficiency Utility Energy Efficiency Programs*. Stanford, CA. Retrieved from http://web.stanford.edu/group/peec/cgi-bin/docs/behavior/research/When Not Losing is Better than Winning-ACEEE Summer Study.pdf.

Thaler, R. (1980). Toward a positive theory of consumer choice. *Journal of Economic Behavior and Organization L*, 39–60. Retrieved from http://citeseerx.ist.psu.edu/viewdoc/download?doi=10.1.1.454.6386&rep=rep1&type=pdf.

UNEP. (2011). *Visions for Change: Recommendations for Effective Policies on Sustainable Lifestyles*. Paris, France. Retrieved from http://www.unep.fr/shared/publications/pdf/DTIx1321xPA-VisionsFor Change report.pdf.

Van Der Horst, D. (2007). NIMBY or not? Exploring the relevance of location and the politics of voiced opinions in renewable energy siting controversies. *Enrgy Policy*, *35*, 2705–2714. http://doi.org/10.1016/j.enpol.2006.12.012.

van Stralen, M. M., de Meij, J., te Velde, S. J., van der Wal, M. F., van Mechelen, W., Knol, D. L., & Chinapaw, M. J. (2012). Mediators of the effect of the JUMP-in intervention on physical activity and sedentary behavior in Dutch primary schoolchildren from disadvantaged neighborhoods. *International Journal of Behavioral Nutrition and Physical Activity*, *9*(1), 131. http://doi.org/10.1186/1479-5868-9-131.

Verplanken, B., & Wood, W. (2006). Interventions to Break and Create Consumer Habits. *Source Journal of Public Policy & Marketing*, *25*(1), 90–103. Retrieved from http://www.jstor.org/stable/30000528.

Werch, C. E. C., Bian, H., Moore, M. J., Ames, S., DiClemente, C. C., & Weiler, R. M. (2007). Brief multiple behavior interventions in a college student health care clinic. *The Journal of Adolescent Health : Official Publication of the Society for Adolescent Medicine*, *41*(6), 577–85. http://doi.org/10.1016/j.jadohealth.2007.06.003.

Wilson, D. K. (2015). Behavior Matters: The Relevance, Impact, and Reach of Behavioral Medicine. *Annals of Behavioral Medicine*, *49*(1), 40–48. http://doi.org/10.1007/s12160-014-9672-1.

Wilson, D. K., Griffin, S., Saunders, R. P., Kitzman-Ulrich, H., Meyers, D. C., & Mansard, L. (2009). Using process evaluation for program improvement in dose, fidelity and reach: the ACT trial experience. *International Journal of Behavioral Nutrition and Physical Activity*, *6*(1), 79. http://doi.org/10.1186/1479-5868-6-79.

Wright, M. (2017). Norwegian school installs tunable LED lighting to enhance learning. *LEDs Magazine*, *July/Augus.*.

Zarrett, N., Skiles, B., Wilson, D. K., & McClintock, L. (2012). A qualitative study of staff's perspectives on implementing an after school program promoting youth physical activity. *Evaluation and Program Planning*, *35*, 417–426. Retrieved from https://ac-els-cdn-com.colorado.idm.oclc.org/S0149718911001170/1-s2.0-S01497189 11001170-main.pdf?_tid=1ed865f4-d5f9-11e7-aa62-00000aab0f6b& acdnat=1512065340_a8695cf47f6cbe64c03d7f4dfcd4e0f3.

In: Energy Policy
Editor: K. J. Sreekanth

ISBN: 978-1-53613-744-6

Chapter 5

THE NIGERIAN ELECTRICITY MARKET: IS THE FUTURE BRIGHT?

Karen Maguire[1,*]***, PhD and Kolawole Olaniyi***[2]
[1]Department of Economics and Legal Studies in Business, Spears School of Business, Oklahoma State University, Stillwater, OK, US
[2]2017 Mandela Washington Fellow, Young African Leaders Initiative (YALI), Ota, Ogun State, Nigeria

ABSTRACT

There are two divergent views of the future of Nigeria's electricity supply, one focused on the lack of investment in grid-based electricity supply, which is leaving Nigerians without power. The second bursting with optimism that distributed renewable generation will solve electricity supply issues throughout sub-Saharan Africa. These two visions of the future of electricity in Nigeria provide a glimpse into the challenges of resolving complex infrastructure issues in developing regions. The optimistic view is based on Nigeria's renewable energy potential, while the pessimistic view has forefront in its mind the difficult challenges of

* Corresponding Author Email: karen.maguire@okstate.edu.

finding sufficient capital investment to channel that renewable potential into powering the homes and businesses of Nigerians. In addition, historically Nigerians have had unreliable access to electricity at best and it is not clear that distributed generation will provide reliable access either. Developed nations today rely on grid-based power systems, in large part because of the redundancy and therefore reliability that they provide. Large scale adoption of a reliable distributed renewable system would leapfrog Nigeria and other sub-Saharan African countries into a new technological future, if it is possible.

1. INTRODUCTION

As of 2016, Nigeria had the largest population and economy on the African continent; it had a population of approximately 186 million and real GDP of US$405.10 billion (WB 2017a, Mungai 2015).[3,4] Its large economy and population, however, was not matched by a similarly sized electricity generation system. It is estimated that the overall demand for electricity generating capacity in Nigeria is as high as 160,000 MW, but the infrastructure is wholly incapable of providing that; the installed generating capacity as of 2015 was 12,522 MW and the transmission capability of the grid was limited to a maximum of 5,300 MWh at a given time (DCS 2016, NIPC 2016).[5] In 2015, the peak generation was only 4,810 MWh, and it was estimated that only 25 percent of generated electricity reached the end user due to losses from an inadequate transmission grid. In addition to its insufficient supply of electricity, the

[3] GDP stands for Gross Domestic Product. It is the "total market value of the goods and services produced by a country's economy during a specified period of time" (Bondarenko 2017). Real GDP is adjusted for inflation (Cambridge Dictionary 2017).

[4] This is an increase from approximately 183,000 in 2015. The real GDP for Nigeria reached a peak of US$568.50 billion in 2014 (WB 2017a).

[5] The 12,522 MW of generating capacity was composed of 10,592 MW of natural gas capacity and 1,930 MW of hydroelectric capacity in 2015 (DCS 2016). Despite significant natural gas resources, over 5 billion cubic meters in proved reserves as of 2016, the transmission capabilities for natural gas are inconsistent and limited (Jannah 2017, OPEC 2017). The unreliable supply of natural gas is cited as one of the factors limiting Nigeria's electricity generation (NPBR 2017, p. 17).

system is also unreliable (Bagu et al. 2016, p. 6, NIPC 2016, NPBR 2016, p. 6).[6]

Unlike the electricity transmission grid in developed countries such as the United States, in Nigeria the transmission grid is limited in its reliability due to a lack of built in redundancies to prevent power outages (DCS 2016).[7] In 2016, Nigerian firms reported that they faced on average eight hours a day of power outages (Bagu et al. 2016, p. 27) In the same year, 96 percent of the electricity consumed by industry was produced on-site primarily through the use of diesel generators rather than through a centralized grid-based system (Bagu et al. 2016, p. 28). According to Frank Jacobs, the President of the Manufacturers Association of Nigeria (MAN), the lack of reliable grid-based electricity has had detrimental effects across the manufacturing industry; including production interruptions, loss of production, and increased unemployment (Yusuf 2016, Oketola 2015). Use of generators is costly and very unreliable due to the fact that diesel fuel costs are significantly higher per kilowatt-hour (kWh) than grid-based electricity and diesel fuel and natural gas are often unavailable (Bagu et al. 2016, p. 21, NPBR 2016, p. 6).[8] Jacobs argues that MAN member firms spent N20.8 billion per month in 2015 and 2016 to generate their own power due to the lack of reliable grid-based electricity (Yusuf 2016).

For small grid-connected Nigerian businesses, the lack of a sufficient and reliable supply of electricity has led to disruptions in existing business activity that have hampered Nigeria's overall economic growth. For example, on April 10, 2016, the Nation Newspaper featured the story of Tolulope Akani, a shoe cobbler who is now out of the shoe making business. Due to the limited and erratic grid-based power supply, he was using a diesel generator. Unfortunately, a midnight fire ignited by a spark from the generator razed the factory (Yusuf 2016). Another example is the

---

[6] On average only 3,879 MW of capacity was on-line from January to August 2015 (NPBR 2016).

[7] Instead, the grid infrastructure is limited to "a single path of transmission with a power source at one end" (DCS 2016).

[8] The cost of diesel fuel in 2015 ranged from NGN 62-94 per kWh while the cost of grid-based electricity ranged from only NGN 26-38 per kWh (NPBR 2016, p. 6).

story of Bimbola Akinyemi featured in *PUNCH*. She began a poultry business selling frozen poultry products in 2013, but her products repeatedly spoiled due to power outages. She relied on a backup generator, but she was unable to profitably run the business due to the high cost of fuel for the generator (Oketola 2015).

The constraint on economic development is even more pronounced for those individuals and businesses without access to grid-based electricity. According to the Nigerian Baseline Power Report (NPBR), as of August 2015, approximately 55 percent of the population is not connected to the grid, and therefore does not have access to grid-based electricity (NPBR 2016, p. 4).[9] There were approximately 95 million people out of an estimated 182 million who were not connected to an electricity grid as of 2015 (Bagu et al. 2016, p. 11, NPBR 2016, p. 5). This was in stark contrast to peer African countries, for Ghana the rate was 28 percent and for South Africa 15 percent as of 2016 (NPBR 2016, p. 7). For Nigerians without grid access, their choices are limited; 86 percent of total energy consumption came from biomass in 2016 and many individuals and firms rely on diesel generators for electricity generation (Bagu et al. 2016, p. 6).

Due to the limited generation capacity and the insufficient transmission capability of the current grid system, some have argued that the future of electricity generation in Nigeria must include distributed electricity generation (DCS 2016). A distributed electricity generation system is designed such that electricity is generated and consumed on-site, rather than through a grid-connected system.[10] In this chapter we will examine the recent changes in the structure of Nigeria's grid-based electricity market through the implementation of deregulation and discuss possibilities for Nigeria's electricity future, including the role of both modern renewable generation and a distributed electricity system.

[9] Access is even more restricted in rural areas; only 35 percent of the population has access to grid-based electricity (DCS 2016).

[10] In the United States, an example of a simple distributed generation system is solar panels on a rooftop.

2. THE CURRENT STATE OF NIGERIA'S ELECTRICITY SECTOR

In 2013, Nigeria completed deregulation of their electricity generation and distribution systems (PSRP 2017).[11] Prior to deregulation, the electricity system was federally government owned and managed. The goal of deregulation was to improve the overall system by introducing private firms and competition.

As of 2015, in Nigeria, the electricity system was based on hydroelectric generation, 15 percent, and natural gas generation, 85 percent, both of which require relatively large fixed costs to build the generation and transmission infrastructure, and relatively low costs for fuel and maintenance (NPBR 2016, p. 11; DCS 2016).[12] This leads to the natural monopoly market structure; it is less costly on average for one firm to supply the entire market than it is for multiple firms. The monopoly market structure, however, results in higher prices for consumers than a competitive market (Hubbard and O'Brien 2016, pp. 498-523). Grid-based electricity markets are in general high fixed cost industries, due to the large capital costs of building transmission lines and electricity generating capacity for traditional fuels, i.e., coal, hydroelectric, etc.[13] The natural monopoly market structure is common in the developed world for grid-based electricity as well. In the United States, in a regulated electricity market, the electricity is provided by either a government owned provider, a regulated private company, or an electric cooperative. If a private for-profit company provides electricity generation, it must gain approval from

---

[11] The restructuring began in 2001, followed by legislative changes in 2005, 2007, and 2010 (PSRP 2017, p. 8).

[12] Total costs are assumed to be the sum of fixed costs and variable costs (Hubbard and O'Brien 2016, pp. 367-368). Fixed costs do not change with the amount of electricity generated and include all up-front capital costs. Variable costs depend on the amount of electricity generated and include things such as fuel costs, maintenance, and labor costs.

[13] A natural gas generator typically has lower cost and smaller capacity than a coal or hydroelectric generator, but there are still relatively high fixed costs for construction of natural gas generation as compared with fuel and other variable costs. Also, the large-scale transmission infrastructure is still required so that there is a natural monopoly market structure with large scale grid-based natural gas generation.

a government regulator in order to increase prices, while government and not-for-profit entities are by definition not incentivized to raise prices in order to increase profits.[14]

Nigeria's goal in deregulating its market was to promote competition and improve the supply of electricity generation (PSRP 2017, p. 9). Deregulation was pursued in other parts of the world as well, in the United States, deregulation was a popular policy solution that was adopted by several states in the 1990s after the passage of the Energy Policy Act of 1992 with mixed success; fifteen states have some degree of deregulation in their electricity markets as of 2017 (SOE 2017). The long-term goal was that new firms in a more competitive market would lower prices, adopt new technologies, and promote alternative electricity generating sources such as modern renewables, wind and solar. Generally, in deregulated markets, there is still one set of transmission lines, but electricity suppliers sell electricity into a wholesale electricity market. Other firms then purchase the electricity in the market and sell to consumers. This is designed to promote competition by allowing new electricity generating Independent Power Producers (IPPs) to enter the market to produce and sell wholesale electricity generation and new power distribution companies to enter the wholesale market to purchase electricity and sell to consumers.

In Nigeria in 2010, under the Roadmap for Power Sector Reform Policy, the federal government owned National Electric Power Authority was replaced by the Power Holding Company of Nigeria, which was then divided up into generation and distribution companies that were subsequently privatized in 2013. The transmission system was placed under the management of the Transmission Company of Nigeria, which is owned by the federal government, but as of 2013 is managed by a private company (NPBR 2016, p. 5).[15] Rather than initially implementing a

[14] Private for-profit firms are assumed to be profit maximizing organizations while government and non-profit organizations are by definition not profit maximizing firms.

[15] The government retained ownership of the transmission system (PSRP 2017, p. 8). Centralized ownership and management of the grid is customary in both regulated and deregulated systems due to the natural monopoly characteristics of the electricity transmission grid, i.e., it would not be cost effective to have multiple grids covering the same geographic area.

competitive wholesale market, in order to maintain market stability, deregulation was implemented in stages and, as of 2017, there was a single bulk wholesale electricity trader (NBET) from which the power distribution companies (DisCos) purchased wholesale electricity at prices set by the Nigerian Electricity Regulatory Commission (NERC).[16] NERC establishes wholesale and retail prices in order to cover electricity production costs and incentivize market participants, including DisCos to reduce losses and lower their costs (NERC, 2017a).[17] Deregulation has facilitated the entry of non-governmental entities into the electricity sector allowing for increased competition, which it is hoped will increase generating capacity and reliability. As of 2017, privatization had led to the addition of an IPP and more than US$5 billion in new investments (PSRP 2017, p. 9). The implementation is still struggling to get off the ground, however. In 2017, the grid-based generating capacity had increased to 13,400 MW, of which only 8,000 MW was operational and less than 4,000 MWh per hour was transmitted on average (PSRP, 2017, p. 10). According to Jacobs, MAN President, among several factors leading to a dismal year for manufacturers in Nigeria in 2016 were the "epileptic power supply, bad roads, high interest rate, and high cost of energy" (PT 2017).

In addition to concerns over the generating and transmission capital, serious concerns exist regarding electricity prices and customer billing. Over 50 percent of grid-based electricity customers in Nigeria are not metered, instead receiving estimated bills (NEH 2017; Nnodim 2017; Onyeji 2017). This was an issue prior to deregulation, but it is one of the issues that privatization was implemented to correct (Nnodim 2017).[18] According to NERC, the regulatory body overseeing the generation and distribution of electricity in Nigeria, customers have a right to a meter

[16] NBET is the Nigerian Bulk Electricity Trading Plc. "The bulk buyer or electricity in the Transition Electricity Market" (PSRP 2017, p. 3-4).

[17] As of 2017, the market is in stage 2 of the deregulation process. In the fourth and final stage, the market structure is expected to be fully deregulated with DisCos purchasing electricity from IPPs and GenCos, Power Generation Companies (PSRP 2017, p. 3-4).

[18] The distribution of meters became a problem after a new type of meter was adopted, a pre-paid meter. There have been supply and distribution problems with the new meters and they were not successfully rolled out to the customer base as of 2017 (NEH 2017).

(NERC 2017b). However, the metering of customers has lagged far behind expectations and caused conflict between the DisCos and consumers (Nnodim 2017; Onyeji 2017). Without effective metering, customers may pay more than they actually consume because their bills are estimated. Many consumers are disillusioned by the pace of metering implementation and some are convinced that the DisCos continue to estimate bills in order to increase their revenue by deliberating inflating estimated usage in order to overbill customers (NEH 2017).

Electricity cost concerns for consumers extend beyond estimated bills. There is also conflict regarding the retail electricity rates set by NERC. NERC pricing is set using a Multi-Year Tariff Order (MYTO) that is updated at most every 6 months (NERC 2017a, Okafor 2017). The incremental increase in the electricity price to consumers has been significant at times. For example, the MYTO that took effect in February of 2016 increased electricity prices in one region from N14/kWh to N23.60/kWh (Opeyemi 2016). At the same time, the financial viability of the DisCos is uncertain because the MYTO has consistently been set based on estimated costs that are far below the DisCos realized costs (PSRP 2017, p. 9).[19] In Nigeria, the MYTO is set based on the ATC&C, the aggregate technical, commercial, and collections loss. The ATC&C is the difference between the value of the amount of electricity received by a DisCo from the wholesale purchaser (NBET) and the revenue that is collected from consumers (IWIN 2016). The ATC&C projected by NERC has consistently significantly understated the true generation and collection losses incurred by the DisCos. In 2016, the MYTO was calculated on a projected loss of 32.1 percent while the realized loss on average was 54.3 percent, leaving DisCos with a significant deficit (PSRP 2017, p. 12).

[19] Financial difficulties in the electricity sector after deregulation are not unique to Nigeria's electricity market. High electricity prices followed by a regulatory intervention freezing retail prices led to a divergence between the wholesale and retail price of electricity in California after deregulation. A dire situation that included power outages and concerns over financial insolvency of retail power providers developed approximately 18 months after California implemented deregulation in March of 1998, resulting in the roll back and suspension of major portions of the deregulation policy in September of 2001 (US-EIA 2017).

Another factor that led to the financial instability of the system was that the DisCos were significantly behind on their payments to the wholesale purchaser (NBET), collecting 57 percent of customer payments due and remitting only 29 percent to NBET in 2016 (PSRP 2017, p. 28). Further adjustments in the MYTO that result in increased consumer prices were being proposed in mid-2017 with updates on a quarterly or monthly basis (Okafor 2017). Given the fact that despite the price increases between 2013 to 2017, there had been no significant improvement in the electricity distribution, stability, or availability, higher rates may be a hard sell for the DisCos and NERC (NPBR 2016, p. 30).

On the whole, the electricity sector has not had a consistent improvement in generation or transmission and several factors must continue to be addressed. These include government interventions to maintain the financial viability of electricity generators and distributors, improving the transmission of natural gas to electricity generators, and improving the transmission infrastructure so that it can transport the existing electricity generation supplied to the grid (DCS 2016; NPBR 2016, p. 30). In 2017, the Nigerian government launched the Power Sector Recovery Program in order to improve the reliability of the grid-based infrastructure and stabilize the financial position of the firms operating in the deregulated market (WB 2017b). The Nigerian government continues to recognize that a sufficient source of reliable energy is vital for economic growth in the country (WB 2017b). Given the existing constraints in the grid-based system, some have begun to propose that a viable distributed electricity generation system should be implemented in order to supplement the grid-based system (DCS 2016).

3. Distributed Electricity Generation and Renewable Energy

Distributed energy generation has been in existence for decades, pre-dating large grid-based electricity systems, at a time when all consumers were required to meet their energy needs on-site. An example of such is a homesteader in the United States using a small wind turbine to pump water from a well in the late 19th century. Subsequently, when commercial scale electricity generation began, small, local grids were the norm. As electricity generation became widespread in developed countries, both the generating capacity and transmission grids increased in size in order to take advantage of economies of scale (Pepermans et al. 2005).[20] Economies of scale occur when per unit costs are falling as the quantity produced is rising (Hubbard and O'Brien 2016, p. 380).

In the 2010s, the use of distributed systems for modern electricity generation became more popular in developed countries because technology growth lowered the cost of distributed renewable energy sources, such as solar (Barbose and Darghouth 2017, p. 2-4, 13).[21] An example of a basic small-scale distributed electricity system is solar panels on a rooftop. They are installed on-site and disconnected from other users, with each individual relying on their own system to meet their electricity demands. Improved technology and lower costs for small-scale solar panel installations since 2000 have spurred grid-connected consumers in the United States to consider the possibility of supplying a portion of their electricity demand through on-site solar electricity generation (Barbose and Darghouth 2017, p. 13).[22] This allows individuals to have their own distributed solar system to meet their electricity demands while at the same time allowing them access to the reliability provided by remaining connected to the grid.[23] According to the U.S. Energy Information

[20] For example, in the United States, electricity generation was primarily coal-fired generation and coal generating capacity has large economies of scale.

[21] The adoption of distributed electricity through residential and commercial solar panel installation is widespread in developing countries. Prices vary substantially, for instance, the cost for on-site installed solar generation from solar panels in Australia is less than half the estimated median U.S. price per watt (Barbose and Darghouth, p. 2)

[22] In the United States, an estimated 52 percent of small-scale solar PV generation was from residential consumers in 2016 (US-EIA 2016).

[23] There is an on-going debate about the fees that grid-connected utility customers with solar panels should pay for grid access and the prices that utilities are required to pay to homeowners who produce excess electricity and transmit that electricity to the grid. The policies vary by country and by state within the United States and are dynamic. See for example, Warrick (2015) for more information on net metering policies in the United States.

Association (US-EIA), small-scale grid-connected solar systems have grown significantly since the mid-2000s.[24] The US-EIA reports that out of total solar generation in the United States in September 2015, an estimated one-third of generation was from distributed solar systems (US-EIA 2015).

As demand for distributed systems has grown, technology has continued to develop. For example, in Japan, after the 2011 Fukushima nuclear disaster, some have proposed that the restoration of electricity does not require relying entirely on reconstructing the grid-based infrastructure. In an article at Reuters.com, Sheldrick and Tsukimori (2017), describe the use of micro-grids in the Japanese city of Higashi Matsushima. Self-sufficiency and concerns over nuclear generation after the Fukushima nuclear disaster are the key factors inducing this shift to distributed systems. The city lost three-fourths of its homes and significant infrastructure in the 2011 earthquake and tsunami and is planning to build an alternative micro-grid electricity system. As of 2017, the city of 40,000 people is planning to produce 25 percent of its electricity using small locally sited generation capacity connected to micro-grids, a distributed electricity distribution solution that incorporates some of the reliability benefits of a large-scale grid-based system. Rather than being fully dependent on a single on-site electricity generation source, such as a solar panel on the roof of a home, the system will recreate the previous grid-based system, but on a much smaller municipal scale. The electricity capacity used will not rely on large-scale nuclear generating plants, but rather on locally sited capacity from natural gas, solar, and wind generated electricity (Sheldrick and Tsukimori 2017).

The success or failure of a new distributed model of electricity generation will depend largely on the costs of the new small-scale generation and micro-grid transmission systems per kWh of electricity produced as compared to traditional large-scale generation and transmission systems. In a traditional electricity generating system there are economies of scale in electricity generation, that lead to lower per kWh

---

[24] According to the US-EIA, small-scale solar systems are defined as those that have less than one MW of generating capacity (US-EIA 2016).

costs if more electricity is produced. This is true of traditional sources of large-scale electricity generating capacity, such as coal, nuclear, and large scale hydroelectric capacity. The costs of constructing the plant are extremely large, high fixed costs, as compared to the per-kWh costs of generating electricity. Consequently, for traditional fuel sources localized capacity and micro-grid distribution would be costlier and lead to higher electricity prices for consumers as compared to a traditional large-scale generation and transmission system. In the Japanese case, however, the fuel sources that they are implementing, natural gas, wind, and solar, have lower fixed costs and therefore a smaller system will generate a sufficient quantity to take advantage of the economies of scale. Also, there are economies of scale in the transmission infrastructure as well, but by building a small scale infrastructure that services an entire community they may be able to build the micro-grid to a sufficient size that they can take advantage of the economies of scale in transmission as well.

In Bangladesh, there is an example of a successful small-scale solar program that has been implemented for some rural households (DCS 2016). The goal of the government supported program is to provide a renewable electricity generating source for rural Bangladeshis that do not have grid access. The Infrastructure Development Company Limited (IDCOL) was established in Bangladesh in 1998 to promote financing for renewable development (IDCOL 2014a). In 2003, the Solar Home System (SHS) Program was implemented to provide off-grid household solar panels for rural Bangladeshis (IDCOL 2014b). The program has led to an increase in off-grid solar generated capacity from zero to 181.7 MW between 2003 and 2016 (IRENA 2017). Given that only 35 percent of the rural Nigerian population has access to grid-based electricity, a similar program could provide a source of electricity for many Nigerians that currently have no access to it (DCS 2016).

In fact, as of 2017, foreign investment in micro-grid solar technology began in Nigeria and the product sold in Nigeria is similar to the solar panels provided in Bangladesh. (Hanley 2017; Shalev 2017). Lumos, a firm based in the Netherlands expanded their household solar panel kit into Nigeria in 2017. The kit is a single rooftop solar panel and battery meant to provide enough electricity to provide lighting, charge mobile phones, and run small household appliances. The product has already been successfully implemented in Kenya, Rwanda, and India. The solar panel kits are sold at an initial fixed cost with a per kWh usage fee that is pre-paid by households. As of March 2017, the company had sold 30,000 solar panel kits in Nigeria and had plans for a significant increase to 10 million kits in five years (Shalev 2017). As this example demonstrates, the future of the electricity market in Nigeria may be in distributed generation if it can be provided in a cost-effective way.

According to estimates from the International Renewable Energy Agency (IRENA), however, between 2012 and 2016, Nigeria added a total of 2.2 MW of off-grid solar PV capacity, and between 2002 and 2014, 3.2 MW of wind capacity (IRENA 2017). This led to a total of 17.2 MW of distributed solar capacity and a total of 3.2 MW of distributed wind capacity in 2016 (IRENA 2017). The numbers are minute and the distributed renewable potential in Nigeria remains largely untapped. Still, as of 2017, the inadequate state of grid-based electricity distribution and its limited reach particularly in rural areas provide fertile ground for off-grid solutions, both from renewable energy such as solar and non-renewable sources.

As of 2016, grid-based renewable capacity in Nigeria was extremely limited, and non-hydroelectric grid-based renewable generation capacity did not exist. According to estimates provided by IRENA, since 2001, the only grid-based renewable electricity capacity that had been added to the grid was 100 MW of hydropower in 2012. As of 2015 there were three hydroelectric power plants, each with a capacity of greater than 500 MW and a total of approximately 2,000 MW of hydroelectric generation (NPBR 2016, p. 11-12; IRENA 2017). There were no grid-based solar or wind capacity additions despite Nigeria's large potential for solar generated electricity (IRENA 2017, NERC 2017c).

4. The Future of Nigeria's Electricity Market

Nigeria's continued economic development is dependent on a reliable source of electricity that is of sufficient scale to meet the demands of all sectors of the economy; residential, commercial and industrial. There are significant improvements needed in the capacity and reliability of the grid-based generating and transmission system. The implementation of deregulation and government support for improving the electricity system have increased foreign investment in Nigeria's electricity system, however. In 2017, the first grid-based non-hydroelectric generated capacity was planned. Greenwish Partners, an IPP based in France, plans to invest hundreds of millions of dollars to add 200 MW of solar generating capacity in Nigeria. The CEO, Aubin-Kalaidjian specifically noted the policies of the Nigerian government in promoting growth in the electricity system as one of the reasons to invest in Nigeria (Renner 2017).

Concerns remain and due to the current state of the grid-based generating and transmission system, the implementation of distributed generation systems may provide the most cost-effective way forward for Nigerian households and firms (Ajao et al. 2009). Projects such as those described above from Lumos could provide a source of electricity for many rural and non-grid connected Nigerians that currently have no access to it and there are additional projects planned. A joint project between Nigerian firm, Community Energy Solar Enterprises Limited (CESEL), and American firm, Renewvia Energy Corporation, is planned to provide 10 MW of distributed solar generation across twenty-five Nigerian communities by the end of 2017. The goal is to provide sufficient affordable electricity for LED lighting and cell phone charging for households and street lights in communities (Hanley 2017).[25] In 2016, NERC approved regulations allowing distribution companies to use micro-grids. The goal is for these micro-grids to expand access to rural areas and

[25] The project was in part financially supported by USAID's Power Africa project (Hanley 2017), and is not yet clear if the project will be financially viable over the long term. USAID is a "U.S. Government agency that works to end extreme global poverty and enable resilient, democratic societies to realize their potential" (USAID 2017).

areas which have limited or no access to large-scale grid-based electricity (Ola 2016).[26]

Distributed capacity is not limited to small-scale rural generation with renewables, however. Nigeria's industrial users are also adopting small-scale urban natural gas or other generated electricity capacity with micro-grids. NERC has encouraged industrial users in urban areas to adopt this model. The goal is to improve the reliability and amount of generation available for industrial users (NERC 2015a). As of 2015, the Nigerian government was working with MAN to develop a regulatory framework that would allow industrial users who generate excess electricity on-site to sell it to neighboring firms, create rules for the establishment of local small-scale generation and micro-grids, and provide a regulatory framework to allow the industrial micro-grids to connect to the existing grid system (NERC 2015a; NERC 2015b). In addition to the direct benefits to industrial consumers, the hope is that the micro-grids will benefit residential and commercial electricity consumers indirectly through increased access to existing grid-based generation (NERC 2015b). The long term goal is of course to increase economic development, benefiting both firms and consumers.

Nigeria's future economic development relies on reliable, scalable electricity generation. The current path forward for the country is to both improve the grid-based infrastructure and supply of electricity and at the same time pursue both renewable and non-renewable distributed generation. It remains to be seen which distribution method will become dominant. Factors including costs per kWh, scalability, and reliability will all play a role. As both micro-grid and renewable technology improve and costs decline, the future of electricity generation for Nigeria surely lies with a combination of grid and off-grid technologies.

---

[26] In the Nigerian regulation, micro or mini-grids are defined as having an electric capacity of 10kW or larger. Generating systems with a capacity of 100kW to 1 MW will require a government permit, while smaller systems can be operated without one (Ola 2016).

5. Conclusion

As of 2017, the Nigerian electricity generation system cannot meet the needs of consumers. An estimated demand of 160,000 MW of capacity greatly outstrips the existing generating capacity of 13,400 MW and the available generating capacity of 8000 MW as of 2017 (PSRP, 2017, p. 10; DCS 2016, NIPC 2016). Deregulation, while still in its early stages, has so far been unable to significantly improve the supply and reliability of the grid-based electricity system and a large renewable potential largely remains untapped with the exception of existing hydroelectric capacity and some planned grid-based solar capacity (IRENA 2017; NPBR 2016, p. 30). The potential of distributed electricity generation is in its infancy, but technology is continuing to evolve that is lowering costs and there are examples in other countries, Bangladesh and Japan, that may provide a successful blueprint for distributed Nigerian electricity generation (Barbose and Darghouth 2017; Sheldrick, and Tsukimori 2017; IDCOL 2014b). The Nigerian government through NERC has already begun to promote distributed generation and micro-grids for both rural non-grid connected households and urban industrial users and private firms have begun to implement distributed solar generation projects (Ola 2016, NERC 2015a). Still, the negative economic impacts of insufficient and unreliable electricity generation are felt across all sectors of the economy. The path forward remains unclear, but the Nigerian government recognizes that a sufficient source of reliable electricity is vital for economic growth in the country and continues to pursue improvements to the deregulated grid-based electricity market, while at the same time promoting distributed electricity policies (WB 2017b). It will take a combination of on-grid, off-grid, traditional, and renewable fuels to provide the electricity that Nigerians need to power their economy.

REFERENCES

Ajao, K. R., Ajimotokan, H. A., Popoola, O. T. & Akande, H. F. (2009). Electric energy supply in Nigeria, decentralized energy approach. *Cogeneration and Distributed Generation Journal*, *24*(4), pp. 34-50.

Bagu, Thessa., Thomas, Dietz., Emiel, Hanekamp., Aniche, Phil-Ebosie. & Bolade, A. Soremekun. *Africa-EU Renewable Energy Cooperation Programme - Captive Power in Nigeria: A Comprehensive Guide to Project Development.* Ed. Ina de Visser. Eschborn, Germany: European Union Energy Initiative Partnership Dialogue Facility, 2016.

Barbose, Galen. & Naïm, Darghouth. (2017). *Tracking the Sun 10: The Installed Price of Residential and Non-Residential Photovoltaic Systems in the United States.* Berkeley, CA: Energy Technologies Area, Lawrence Berkeley National Laboratory.

Bondarenko, Peter. (2017). Gross Domestic Product. *Encyclopedia Britannica.* https:// www.britannica.com/topic/ gross-domestic-product (accessed 10/23/17).

Cambridge Dictionary. (2017). Real GDP. http://dictionary.cambridge. org/us/dictionary/english/real-gdp (accessed 10/23/17).

DCS. (2016). Opportunities for off-grid solutions in the Nigerian power sector. *Detail Commercial Solicitors.* http://www.financialnigeria. com/opportunities-for-off-grid-solutions-in-the-nigerian-power-sector-sustainable-293.html (accessed 10/11/17).

Hanley, Steve. (2017). 10,000 Nigerian Homes to Get Electricity from Solar Microgrids. *CleanTechnica.* https://cleantechnica.com/ 2017/03/ 02/10000-nigerian-homes-get-electricity-solar-microgrids/ (accessed 10/27/17).

Hubbard, R. Glenn. & Anthony, Patrick O'Brien. (2016). *Microeconomics*, 6th Edition. Boston, MA: Pearson, pp. 367-368, *380*, 498-523.

IDCOL. (2014a). About IDCOL. *Infrastructure Development Company Limited.* http://idcol.org/home/about (accessed 10/21/17).

IDCOL. (2014b). Renewable Energy: Solar Home System Program. *Infrastructure Development Company Limited.* http://idcol.org/home/ solar (accessed 10/21/17).

IRENA. (2017). *Capacity and Generation. International Renewable Energy Agency.* http:// resourceirena.irena.org/ gateway/ dashboard/ ?topic=4 &subTopic=19NGA (accessed 10/20/17).

Jannah, Chijioke. Why poor electricity is still a problem in Nigeria – Power Minister, Fashola. *Daily Post – Nigeria News.* http://dailypost.ng/2017/05/28/poor-electricity-still-problem-nigeria-power-minister-fashola/ (accessed 10/20/17).

IWIN. (2016). The Nigerian Electricity tariff and the ATC&C loss paradox: are we paying for services or losses. *Independent Energy Watch Initiative.* http://iwin.org.ng/index.php/news/item/2235-the-nigerian-electricity-tariff-and-the-atc-c-loss-paradox-are-we-paying-for-services-or-losses (accessed 10/13/17).

Mungai, Christine. (2015). How will Africa change by 2030? World Economic Forum. https://www.weforum.org/agenda/2015/06/how-will-africa-change-by-2030/ (accessed 10/20/17).

NEH. (2017). DisCos, Customers and the Metering Conundrum. Nigeria Electricity Hub. http://www.nigeriaelectricityhub.com/2017/06/ 21/ discos-customers-and-the-metering-conundrum/ (accessed 10/13/17).

NERC. (2017a). Electricity Tariff in Nigerian Electricity Supply Industry. Nigerian Electricity Regulatory Commission. http://www.nercng.org/ index.php/home/myto (accessed 10/13/17).

NERC. (2017b). Consumer Rights & Obligations. Nigerian Electricity Regulatory Commission. http://www.nercng.org/index.php/home/ consumers/consumer-rights-obligations (accessed 10/13/17).

NERC. (2017c). Renewable Energy Sourced Electricity. Nigerian Electricity Regulatory Commission. http://www.nercng.org/index.php/ home/operators/renewable-energy (accessed 10/20/17).

NERC. (2015a). NERC, MAN Collaborate On Improved Power Supply to Industries. Nigerian Electricity Regulatory Commission. http:// www.nercng.org/index.php/media-library/press-releases/309-nerc-man-collaborate-on-improved-power-supply-to-industries (accessed 10/21/ 17).

NERC. (2015b). NERC & MAN to Develop Framework to Expedite Electricity Supply to Industries. Nigerian Electricity Regulatory

Commission. http://www.nercng.org/index.php/media-library/press-releases/293-nerc-man-to-develop-framework-to-expediate-electricity-supply-to-industries (accessed 10/21/17).

NIPC. (2016). Invest in Nigeria: Explore the Possibilities. Nigerian Investment Promotion Commission. http://www.invest-nigeria.com/power/ (accessed 10/11/17).

Nnodim, Okechuwu. 2.9 million power consumers still unmetered, says NERC. PunchNG.com. http://punchng.com/2-9-million-power-consumers- still-unmetered-says-nerc/ (accessed 10/12/17).

NPBR. (2016). Nigerian Power Baseline Report. Nigerian Electricity Supply Industry Stats. http://nesistats.org/briefings.html (accessed 10/11/17).

Okafor, Chineme. (2017). NERC mulls monthly review of electricity tariffs. THISDAY Newspapers LTD. https://www.thisdaylive.com/index.php/2017/06/08/nerc-mulls-monthly-review-of-electricity-tariffs(accessed 10/27/17).

Oketola, Dayo. (2015). Nigerian small businesses on 'death row' – SMES, traders losing investments to power shortage. PunchNG.com. http://punchng.com/nigerian-small-businesses-on-death-row-smes-traders-losing-investments-to-power-shortage/ (accessed 10/18/17)

Ola, Danielle. (2016). Nigeria Electricity Regulation Commission approves mini-grid licenses. PV-tech.org. https://www.pv-tech.org/news/nigeria-electricity-regulation-commission-approves-mini-grid-licences (accessed 10/21/17).

Onyeji, Ebuka. (2017). Electricity regulator, NERC, directs Nigerians without prepaid meters to stop paying bills. Premium Times. https://www.premiumtimesng.com/news/top-news/233653-electricity-regulator-nerc-directs-nigerians-without-prepaid-meters-to-stop-paying-bills.html (accessed 10/13/17).

OPEC. (2017). Nigeria facts and figures. Organization of Petroleum Exporting Countries. http:// www.opec.org/opec_web/en/ about_us/167.htm (accessed 10/20/17).

Opeyemi, Damilare. (2016). Here is what you should know about Nigeria's new electricity tariff. Ventures Africa.http://venturesafrica.com/is-the-new-electricity-tariff-justifiabe/(accessed 10/27/17).

PSRP. (2017). Power Sector Recovery Program: 2017-2021. Federal Republic of Nigeria. http:// www.nesistats.org/ uploads/3/4/9/3/ 34934878/psrp__doc_-_latest_may_9_.pdf (accessed 10/12/17).

PT. (2017). Nigerian manufacturers lament: 272 first shut, 20% capacity utilization, other challenges in 2016. Premium Times. https:// www. premiumtimesng.com/news/headlines/220035-nigerian-manufacturers-lament-272-firms-shut-20-capacity-utilisation-challenges-2016.html (accessed 10/17/17).

Renner, Solape. (2017). GreenWish to Invest $280 Million in Nigeria Solar Plants. RenewableEnergyWorld.com. http:// www.renewableenergy world.com/articles/2017/03/greenwish-to-invest-280-million-in-nigeria-solar-plants.html (accessed 10/27/17).

Shalev, Asaf. (2017). Solar Power Taking Hold in Nigeria, One Mobile Phone at a Time. InsideClimate News. https:// insideclimatenews.org/ news/07022017/solar-energy-nigeria-africa-renewable-energy-climate-change-lumos (accessed 10/27/17).

Sheldrick, Aaron. & Osamu, Tsukimori. (2017). Quiet energy revolution underway in Japan as dozens of towns go off the grid. Reuters.com. https://www.reuters.com/article/us-japan-energy-revolution/quiet-energy-revolution-underway-in-japan-as-dozens-of-towns-go-off-the-grid-idUSKCN1BU0UT (accessed 10/20/17).

SOE. (2017). Energy Deregulation by State. SaveOnEnergy.com. https:// www.saveonenergy.com/state-information/ (accessed 10/13/17).

USAID. (2017). What We Do. USAID. https://www.usaid.gov/what-we-do (accessed 10/27/17).

US-EIA. (2017). Subsequent Events-California's Energy Crisis. Energy Information Administration. https://www.eia.gov/electricity/policies/ legislation/california/subsequentevents.html (accessed 10/13/17).

US-EIA. (2016). More than half of small-scale photovoltaic generation comes from residential rooftops. U.S. Energy Information Association.

https:// www.eia.gov/ todayinenergy/ detail.php?id=31452 (accessed 10/21/17).

US-EIA. (2015). EIA electricity data now include estimated small-scale solar PV capacity and generation. U.S. Energy Information Association. https://www.eia.gov/todayinenergy/detail.php?id=23972 (accessed 10/18/17).

Warrick, Joby. (2015). Utilities wage campaign against rooftop solar. Washington Post. https://www.washingtonpost.com/national/health-science/utilities-sensing-threat-put-squeeze-on-booming-solar-roof-industry/2015/03/07/2d916f88-c1c9-11e4-ad5c-3b8ce89f1b89_story.html?utm_term=.6b29209554a4 (accessed 10/20/17).

WB. (2016). DataBank: World Development Indicators. *The World Bank.* http://databank.worldbank.org/data/reports.aspx?source=2&country=NGA&series=&period= (accessed 10/11/17).

WB. (2017a). The World Bank: Data – Nigeria. *The World Bank.* https://data.worldbank.org/country/nigeria (accessed 10/23/17).

WB. (2017b). World Bank Group Support for Nigeria's Power Sector Recovery Program. *The World Bank.* http:// www.worldbank.org/en/news/press-release/2017/04/22/world-bank-group-support-for-nigerias-power-sector-recovery-program (accessed 10/13/17).

Yusuf, Ibrahim Apekhade. (2016). Businesses bear the brunt of unstable electricity. The Nation. http://thenationonlineng.net/businesses-bear-brunt-unstable-electricity/ (accessed 10/17/17).

In: Energy Policy
Editor: K. J. Sreekanth

ISBN: 978-1-53613-744-6

Chapter 6

INSTITUTIONALISATION OF THE COMMON GAS MARKET IN THE CONTEXT OF INSTITUTIONAL EVOLUTION OF THE EURASIAN ECONOMIC UNION

***Elena Shadrina**, PhD**
School of International Liberal Studies,
Waseda University, Tokyo, Japan

ABSTRACT

The chapter addresses some of the aspects of the ongoing institutionalisation in the Eurasian Economic Union (EAEU). Having briefly overviewed the history of the EAEU formation, it describes the overall institutional environment of the EAEU and characterises the contemporary stage of the EAEU's economic integration depicting the impediments and the opportunities for the latter. The rest of the chapter is devoted to exploring the common gas market (CGM) recently endorsed by the institutional provisions of the EAEU. In doing so, it analyses contemporary gas relations within the EAEU with application of the most

* Corresponding Author: Nishi Waseda 1-6-1, Shinjuku-ku, Tokyo, 169-8050;
Email: shadrina@aoni.waseda.jp.

recent data published by the Eurasian Economic Commission, examines the CGM's model and its principal institutions as they are envisioned. This chapter discusses potential benefits and concerns of the economies engaging in the CGM, as well as speculates on possible developments involving third parties. It concretises those through the analysis of the most salient issues of the CGM's formation as they relate to gas price, pricing, competition and others. The chapter argues that despite long existing constraints for integration in the EAEU due to mainly the factor of the member states economies' structural similarity, institutionally compatible economies display their intention to progress beyond supranational institution-building to include sectoral initiatives, within which the studied example of the CGM falls. The chapter demonstrates that the CGM exploits the existing institutional complementarity among the respective national gas sectors and extends Russia's somewhat more advanced institutional arrangements to the projected common gas market.

Keywords: Eurasian Economic Union, Common Gas Market, institutions

ACRONYMS

bcm	billion cubic meters
bn	billion
CBR	Central Bank of Russia
CES	Common Economic Spaces
CGM	Common Gas Market
CIS	Commonwealth of Independent States
CU	Customs Union
EADB	Eurasian Development Bank
EAEU	Eurasian Economic Union
EEC	Eurasian Economic Commission
EFSD	Eurasian Fund for Stabilization and Development
EU	European Union
GDP	gross domestic product
mn	million
OBOR	One Belt One Road
PPP	purchasing power parity

1. Introduction

In 2011, the Eurasian Economic Union (EAEU) revealed its plan to establish common energy markets. The scholarly attention to the theme was largely missing due to the overall sluggish progression of integrative initiatives within the Eurasian Economic Space and the prevalent scepticism about the future of EAEU. Recently, the theme of common energy markets has been touched upon in the studies on the Eurasian integration (Pastuhova and Westphal 2016, Dragneva and Wolczuk 2017), but no specific study has been presented on the EAEU's common gas market (CGM) which is a part of three common energy markets initiative.

Overall, recent revival of the EAEU can be most of all associated with Russia's proactive attempts to create new economic opportunities in order to mend its great financial losses and commercial damage caused by the deterioration of economic relations with the West and especially with the EU in the aftermath of the Ukrainian crisis. In a changed geopolitical environment, Russia has been advancing its vision of a Greater Eurasia as opposed to its earlier favoured concept of a United Eurasia spanning from Lisbon to Vladivostok.[27] Despite its rather modest economic scale, the EAEU is being energetically promoted by Russia as one of the projects to help the materialisation of Russia's masterplan for the Greater Eurasia.[28] While this chapter concentrates on exploring institutional shifts in the EAEU's gas sector, it offers some analysis of manifold developments towards integration among those post-Soviet economies, which so far have

[27] When in 2010, Russia stepped forward with the idea of United Eurasia, it was genuinely optimistic about the prospects of close economic integration with the EU. There have been sensible anticipations of the bilateral Russia-EU free-trade zone, greater cooperation in industrial sector, more homogenous educational space, visa-free travelling and so on. Mirroring the drastically changed geopolitical environment, in 2016 Russia announced its vision of Greater Eurasia, which can be interpreted as Russia's turn away from the EU and towards a broad range of actors including China, the post-Soviet Central Asian economies, as well as such major emerging economy as India and important to Russia for a variety of reasons Iran, Pakistan and others.

[28] Art. 21, Clause 2 of Strategiya ekonomicheskoi bezopasnosti Rossiiskoi Federatsii do 2030 goda [Strategy for Economic Security of the Russian Federation to 2030]. Approved by the Presidential Order No. 208 on May 13, 2017.

been less definite in setting their agendas for formal economic integration with the EU.

The Concept for the CGM for the EAEU was adopted in May 2016, nearly a decade ahead of the planned materialisation of the market on January 1, 2025. The CGM will be the final act in the implementation of the EAEU's joint energy markets initiatives. The first market to become integrated within the EAEU is electricity (by 2019) followed by the market for oil and oil products (by 2020). Such sequence is largely rational given the already achieved level of interconnectedness across the EAEU's economies in three respective sectors. The characteristics of the national markets also stipulate the timing for the opening up of the domestic markets and linking them into a common market space. It is more rational to start from the markets where the extant regulatory provisions are more uniform and in line with the market principles (electricity and oil markets) and move on to gradually developing common institutions for the market where the initial regulatory differences are more significant.

This account intends to contribute to the scholarly work on the EAEU integration in several ways. First, it extends analysis of the EAEU beyond the subject of economic integration to incorporate institutional theory drawing upon the notions of institutional complementarity and institutional change. Second, unlike the work referenced in this chapter, this account focuses on a sectoral level by looking at the most distant chronologically integrative initiative of the EAEU – CGM. The chapter advances a vision of the CGM institution building as an evolutionary process induced by the complementarity of institutional environments in the EAEU member states.

The organisation of this chapter is as follows. It first outlines conceptual framework by reviewing the most relevant to this work notions of institutional theory. After the EAEU's institutional evolution is briefly reviewed, the characterisation of the principal factors behind a long and difficult start of the Eurasian economic integration is presented. Next, the state of contemporary gas relations among the EAEU member-states is analysed and recent institutional developments in the CGM formation are examined. The assessment of the progress and prospects for the institutionalisation of the CGM is presented upon the notions of institutions, institutional complementarity and institutional change. Summarising observations on the theme at hand conclude the chapter.

2. The Conceptual Framework

Before analysing the evolution of institutional setting of the EAEU and CGM and assessing the prospects for the latter integrative initiative, it is expedient to clarify the notion of institutions, institutional change and institutional complementarity as applied to this work.

Despite the acknowledged importance of institutional theory in socio-economic studies (Andrews-Speed 2016, Lockwood et al. 2017), there is no established definition of institutions. Several strands of institutionalism – rational choice, historical, social and discursive – are being advanced (Andrews-Speed 2016), but most of academic accounts on this theme draw on critical notions developed by North (1990, 1991, 1994, 1995, 1997). Fundamentally, institutions are viewed as humanly devised (North 1990) durable (Aoki 2001) rules, whose principal merit is in reducing the uncertainty associated with regular human transactions (Calvert 1995, Greif 2006).

Institutions in each domain interact informing an equilibrium (Calvert 1995, Aoki 2007). Yet, when the latter is distorted, institutional changes kick in as a reaction of an institutional setting to disequilibrium and its attempt to restore the pre-existing or achieve a new equilibrium. Institutional changes can be either an outcome of an intended course of actions in a certain domain or can occur as a response to the changes taken place in another domain. Earlier prevailing understanding of institutional changes as overwhelmingly incremental and path dependent (North 1995) is greatly complimented by a vision that contemporary institutional changes are increasingly of path-creating nature (Andrews-Speed 2016, Lockwood et al. 2017). Incremental changes can be a product of layering (new elements are added to existing institutions), conversion (new goals or actors are added to existing institution to achieve a desired property of institution) or drift (changing environment combined with policy inaction results either in a gradual change in an institution or in its neglect), but radical institutional changes create new institutional arrangements that are weakly associated with the existing ones and often challenge the latter. Building parallels with energy sector, nowadays, technology- and innovation-intensive sectors, such as renewable energy and smart energy systems, are clearly demonstrating the virtues of path-creation type of radical institutional changes (Andrews-Speed 2016, Lockwood et al. 2017), while institutional changes in traditional sectors tend to continue evolutionary transitions often displaying the logic of conversion (oil and especially natural gas) or layering (LNG sector).

Institutional complementarity - an aggregate property of sundry institutions characterising their ability to function conflict-free and deliver the outcomes with the minimal ex-ante vis-à-vis ex-post gaps in their performance – is important for understanding the nature of institutional change (Crouch 2010). Complementarity between institutions in different domains within national economy (Aoki (2001) is not sufficient for a globally integrated economy; complementarity of its institutional environment in the international context is equally critical. Turning to energy for an illustration, the problem of institutional discrepancy has been a centrepiece of Russia – EU energy relations since the EU started implementing its Third Energy Package promoting liberalisation and deregulation of energy markets in the EU, whereas Russia continued to reinforce features of state capitalism.

Complementarity can compensate for the deficiency of the elements of the institutional system (Hopner 2005, Kleffner 2003). Such are the instances of national and supranational or national and international regulatory settings, especially in closely integrated institutional environments: the EU's Third Energy Package supplementing regulation within the nation-states. Complementarity originating in similarity of the components of the institutional system (Hall 1999, Hall & Soskice 2009) encourage similar kinds of economic actors to behave in analogous manner thereby reinforcing each other's effects. Thus, increasing returns and positive effects on aggregate economic performance serve the criteria of institutional complementarity (Hall & Soskice 2001, Hall & Ginderich 2009). To some extent, state-owned monopolies operating in energy sectors of the EAEU economies can serve as an example of such complementarity. Complementarity can also be understood as an institutional system's ability to enhance its efficacy through the improved performance of its elements (Aoki 2001, 2007; Amable 2000, 2003; Deeg 2005, 2007). An illustration of such, in a broader context of climate change policy, market-based emissions trading instruments supplement the system of direct environmental regulation to enhance the intended outcomes. Institutional complementarity leads to stability of institutional environment, whereas a lack thereof creates momentum for institutional change. Apparently, the process of institutional change may result in previously stabilising complementarities starting to generate cumulated destabilising effects leading to a substantial alteration or even to a breakup of the existing institutional system (Amable 2016).

3. The Eurasian Economic Union

3.1. EAEU Inception and Evolution

While most experts on the Eurasian integration point at the significance of a factor of identity politics as one of the primary driving forces behind the EAEU formation (Aslund 2013, Dreyer and Nicu 2014, Delcoure 2015, Makocki and Nicu 2016), others emphasise that much of the attraction of the EAEU to its member states is tied to Russia's financial aid in the form of energy subsidies (gas for Armenia, gas and oil for Belarus), subsidised loans (Armenia, Belarus and Kyrgyzstan) or remittances, which have been significant in Kyrgyzstan, but also Armenia throughout the transition period (Tarr 2016). Additionally, the importance of security perspective in the EAEU formation is being uniformly agreed upon for that it reinforces Russia's appeal to the Central Asian members of the EAEU, but even more so to Armenia with its complex range of issues in relation to Nagorno-Karabakh region and its bilateral relations with

Turkey (Delcour 2015, Roberts and Moshes 2016, Tarr 2016). More recently, the EAEU member states' incentives in the integrative initiatives are being analysed in a broader context of Greater Eurasia, where the presence of China becomes a crucial factor changing the regional dynamics from a confrontation-driven setting of Russia with its EAEU vis-à-vis EU with its Eastern Partnership to a more complex configuration promising cooperative interaction (Samokhvalov 2016). The advances in the EAEU institution building have made comparisons with underpinned by supranational institutions EU increasingly common (van der Togt et al. 2015, The Eurasian Economic Union… 2016). At the same time, there is discernible scepticism that the EAEU does not stand comparison with the authentically integrated EU (Delcour 2015, Roberts and Moshes 2016, Tarr 2016).

The idea of a Eurasian union was for the first time formulated by Kazakhstan's President Nursultan Nazarbayev in 1994. At that early stage of independence, having few international connections, Kazakhstan was interested to maintain strong economic and security ties with Russia. In the following years, Kazakhstan has announced and successfully introduced the multi-vector foreign policy to fulfil its ambitions to develop balanced relations with Russia, China, EU, USA, Japan and other nations (Mostafa 2013). Pursuing the very same vision in its own foreign policy, Russia nonetheless had been observing the increasing multi-vector orientation of the former Soviet countries not without concern. Accordingly, Russia intensified its appeal to the Commonwealth of Independent States (CIS) members for the enhanced integration emphasising the economic rationale originating in the remaining complementarity among the post-Soviet economic systems. Yet, the formation of the EAEU is predominately perceived as the realisation of the personal ambition of the Russian President Vladimir Putin to enrich his legacy by restoring Russia's influence among the states of the former Soviet Union and strengthen Russia's geopolitical presence in the Eurasian context (Shumylo-Tapiola 2012, Mostafa 2013, Hartwell 2013, Svarin 2016, Dragneva & Wolczuk 2017).

Among the CIS, Russia, Belarus and Kazakhstan have been the most consistent advocates to the idea of economic integration in the post-Soviet space and the most energetic actors in promoting respective initiatives, including the EAEU. In January 2012, Russia, Belarus and Kazakhstan announced their agreement to form a common economic space. This led to the creation of the EAEU envisaging the free flow of goods, services, labour and capital among its members from January 2015.[29] Armenia joined the EAEU to become the fourth member in 2015. The Kyrgyz Republic (Kyrgyzstan) eventually accessed to the EAEU as the fifth member in August 2015 (Kirkham 2016).

It was not until the year 2010, when the CU was commenced, followed by the year 2012 when the CES was formed, that the supranational institutions of Eurasian integration have started to appear (Dragneva & Wolczuk 2013). The CU addressed the problems of codification of a contradictory and fragmented net of bilateral agreements into a unified legal regime. The decisions of the Eurasian Economic Commission and other bodies were given binding legal status. A dispute resolution mechanism was created in 2010, which made it possible for private businesses to appeal against the actions of their counterparts, as well as decisions taken by CU bodies. In 2011, the legal framework of the CU was supplemented with the provisions concerning Russia's accession to the WTO and by the Treaty on the Functioning of the CU in the Multilateral System, which legally obliged the international agreement signed by the CU to comply with the WTO rules (Dragneva &Wolczuk 2012).

The EAEU has fully established institutional architecture composed of supranational and intergovernmental regulatory and judicial bodies. These include the Supreme Eurasian Economic Council, the Eurasian Intergovernmental Council, the Court of the Eurasian Economic Union

---

[29] In December 1991, following the demise of the USSR, the CIS was created. In 1993, CIS members concluded the Economic Union Treaty; the Free Trade Agreement was signed in 1994; the Eurasian Economic Community (EurAsEC) was launched in 2000; the Eurasian Customs Union (ECU), which implied the common external tariff, was launched in 2010; and in 2012 the Common Economic Space (CES) was established. Since 2015, the CES is replaced by the EAEU.

(based in Minsk) and the Eurasian Economic Commission (EEC) as a permanent regulator of the integration process (based in Moscow). The EAEU has its own financial institutions. Established in January 2006 with a registered capital of $7 bn, the Eurasian Development Bank[30] (EADB, based in Almaty) plays a role of a financing mechanism for the EAEU. In addition, in 2009 the Eurasian Fund for Stabilization and Development (EFSD) (formerly, EURASEC Anti-Crisis Fund) with a registered capital of $8.513 bn, was established. This fund is managed by the EADB.

The second stage of institutionalisation of the Eurasian integration is associated with the progress of the EAEU, which has started to be perceived as a new international actor modelled after the EU (Dragneva &Wolczuk 2016). The EAEU portrays itself as a platform to implement fundamental shifts in the architectures of Europe and Eurasia by structuring relations with the EU, ASEAN and China, including a vast host of prospects opening under the latter's One Belt One Road (OBOR) initiative.[31] In 2016, the EAEU concluded its first free trade agreement (FTA) with Vietnam. In 2017, the FTA negotiations were continued with Iran. Various kinds of cooperative schemes are being negotiated with China, India, Singapore, Israel, Egypt and Serbia. Currently, the potential for deeper economic cooperation with South Korea, New Zealand, Cambodia, Mongolia, Peru and Chile is being assessed. The vision of Greater Eurasian Partnership with the EAEU's central role and involving member-states of the EU, Shanghai Cooperation Organisation (SCO)[32] and ASEAN is being increasingly popularised by the EAEU member-states' officials as a response to China's OBOR.[33]

Formally, the EAEU became an international organization endowed with its own legal personality, which is a dramatic progress over the earlier initiatives. Instilling legal status into the EAEU has been instrumental in promoting the vision that other international bodies, such as the EU, will

---

[30] Tajikistan is the only non-member of the EAEU, which is a party to the EADB.

[31] On May 8, 2015, Russia and China signed declaration on cooperation in implementation of the EAEU and the Silk Road Economic Belt.

[32] After India and Pakistan joined the SCO in 2017, the number of its members reached eight.

[33] See recent publications by the Valdai Discussion Club <http://ru.valdaiclub.com>.

need to engage with the EAEU on an inter-regional basis rather than with individual member states (Gusev & Westphal 2015, Pastuhova & Westphal 2016). Yet, the EAEU's institutional and regulatory quality, as well as the discipline of observing the already adopted rules are often found to be suboptimal (Dragneva & Wolczuk 2017). The EAEU countries are performing relatively poorly in the areas of governance and enterprise restructuring as well as competition policy, which signals about their limited progress in more complex institutional and legal reforms (Shadrina 2017b).

3.2. Determinants of Economic Integration in EAEU

Eurasian economic integration has for long been a realm of rhetoric exercises rather than practical accomplishments. Numerous economic, political and social factors frustrated the integrative attempts. On the economic front, the de-industrialisation aggravated by the disruption of intra-Soviet economic ties and a sharp fall in the national outputs added by the hardships of economic transition (such as the skyrocketing inflation, the collapse of public finances, the cessation of the Soviet-type welfare systems and so on) made newly-established sovereign nations concerned utmost about their physical survival. Consequently, in the early stage of economic transition, genuine integration did not appear to be particularly enticing economically to the Eurasian nations and, therefore, was not prioritised politically. The Eurasian states launched profound legal and market reforms with the financial and technical support of the international organisations, such as the World Bank, IMF, ADB and others, formatted upon the principles of the Washington Consensus. Given the challenges of political disintegration in some post-Soviet states, economic integration across former Soviet space appeared to be difficult to achieve. Additionally, the pursuance of multi-vector foreign policy has become a natural course of the sweeping post-Soviet transformations in the sovereign nation-states. In the social context, with the middle-class nearly non-existent in the early transformation period, the cohesive power of interrelated private business interests - an imperative for the progression of authentic economic integration - was overtly missing. Unfortunately, implemented in a neoliberal fashion privatisation in the post-Soviet economies placed the opportunities almost exclusively with the newly established national economic and political elites, thereby allowing them to pursue their private commercial interests both inside and outside the national boundaries. On the other hand, the externalities of the radical economic transformations in the form of growing income inequality (Shadrina 2017b) and deepening social discontent were not conducive to the integrative initiatives.

Although the EAEU made definite progress since its inception, there are certain aspects continuing to affect the dynamism and efficacy of the Eurasian integration. We discuss the most principal of these factors.

The EAEU's economies are highly disproportional in their scale and this is a problematic aspect for integration. That is, in 2016 Russia accounted for nearly 86.6 per cent of the EAEU's GDP whereas Kazakhstan had 9.1 per cent, Belarus - 3.2 per cent, Armenia's contribution was 0.7 per cent and the remainder of 0.4 per cent belonged to Kyrgyzstan.[34] Such asymmetry increases the vulnerability of the Belarusian, Armenian, Kyrgyz and even Kazakhstani economies in the face of negative spill over effects emitted by the Russian economy.

Accurately speaking, Russian economy has been performing unimpressively since the late 2000s (Figure 1). However, the economic downturn became especially apparent in 2014, the year when Russia actively played towards the disruption of the Ukrainian crisis and when the punitive anti-Russian economic, financial and technological sanctions were launched (Shadrina 2016). A deteriorated international setting has additionally weakened the Russian economy, whose shrinking GDP per

[34] Data in this and the following paragraphs of the section are all for the year 2016 and retrieved from: Evraziiskii ekonomicheskii soyuz v tsifrah. Kratkii statisticheskii sbornik. Evraziiskaya ekonomicheskaya komisssiya. Moskva. 2017 [Eurasian Economic Union. Statistics. Eurasian Economic Commission. Moscow. 2017] <http://eec.eaeunion.org/ru/act/integr_i_makroec/dep_stat/econstat/Documents/Brief_Statistics_Yearbook_2017.pdf> (accessed June 1, 2017).

capita in the PPP and real terms has been eventually surpassed by Kazakhstan, as portrayed by Figures 2 and 3.

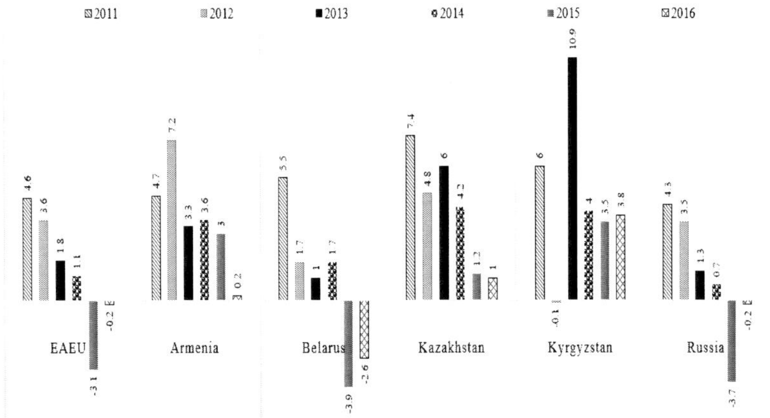

Source: Author based on data http://eec.eaeunion.org/ru/act/integr_i_makroec/ dep_stat/econstat/Pages/statpub.aspx.

Figure 1. Real GDP Growth in EAEU, 2011-2016 (%).

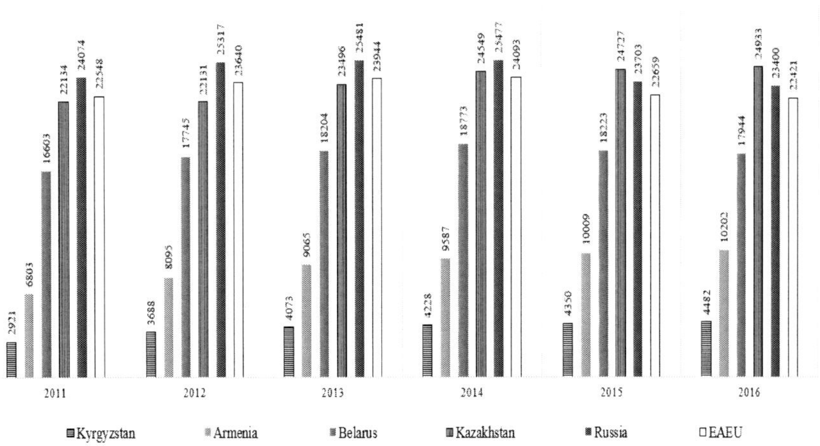

Source: Author based on data http://eec.eaeunion.org/ru/act/integr_i_makroec/ dep_stat/econstat/Pages/statpub.aspx.

Figure 2. GDP per capita PPP in EAEU, 2011-2016 (US$).

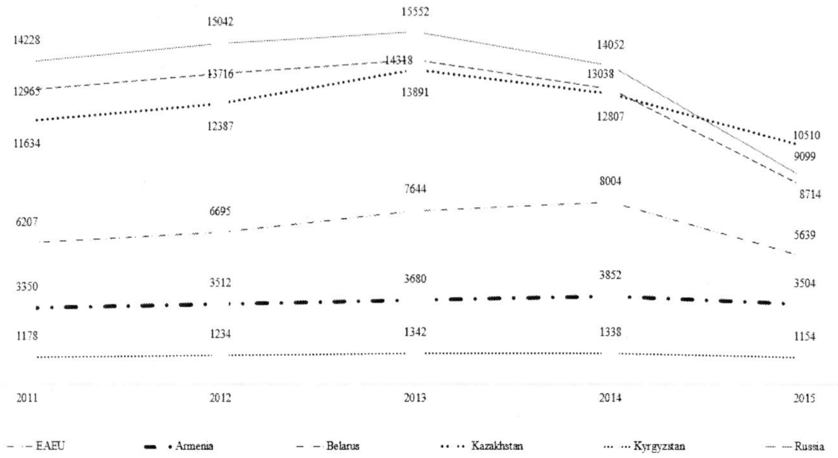

Source: Author based on data http://eec.eaeunion.org/ru/act/integr_i_makroec/dep_stat/econstat/Pages/statpub.aspx.

Figure 3. Real GDP per capita in EAEU, 2011-2015 (US$).

Structural homogeneity of the EAEU economies can equally be named among serious obstacles to integration (Dragneva & Wolczuk 2012, 2013, 2017). Understandably, structurally and technologically similar economies have limited room for cooperation across the value-chain as well as in trade exchange. Such similarity increases the risk of synchronisation of boom-bust cycles between the larger Russian and Kazakhstani economies and smaller EAEU economies. Indeed, Russia and Kazakhstan are fossil fuel dependent economies (Table 1) and Belarus' petrochemical sector is profoundly dependent on Russia for the inputs.

Table 1. Kazakhstan's and Russia's Dependency on Oil and Gas (%)

	Oil and Gas Sector in GDP	Oil and Gas in Merchandised Exports*	Oil and Gas Revenues in Total Fiscal Revenues
Kazakhstan	23	65.08	40-50
Russia	20	59.14	43

Note: (*) data on 2015 export is from Comtrade, HS codes 2709 (crude oil); 2710 (oil products); 271111 (LNG) and 271121 (natural gas);

Source: Adapted from Shadrina (2017b).

Owing to such importance of oil and gas sector to Russia and Kazakhstan, special (protective) provisions for the oil and gas sector of the EAEU will remain in place until 2020. Moreover, the regulations of the Customs Union for oil export tariffs are put outside the EAEU's legal framework. Accordingly, Kazakhstani oil exporters pay higher Russian transit tariffs than domestic companies and their access to the Russian gas pipeline network is restricted.

Overall, it is estimated that Russia is likely to endure heavy costs of integration in the oil and gas sector, which are believed to overweigh the associated benefits. By contrast, Belarus with its developed petrochemical sector is named as the largest beneficiary of the EAEU's energy integration. At the early stage of Eurasian integration, in 2004, the World Bank assessed the value of Russia's gas subsidies alone to Belarus at 7-8 per cent of the country's GDP (Manenok 2016). In 2013, the export price of Russian gas for Belarus was \$173/tcm, which was lower than the average price of gas exported to Europe (\$422/tcm). Such a practice continues: Belarus pays for Russian gas price equivalent to that in Russia's domestic market (with only transportation costs added). In the oil sector, in 2010 Russia imposed tariffs on oil to Belarus on amounts of oil that exceeded Russia's estimate of domestic use in Belarus, which dramatically reduced Belarus' revenues on refined oil (Tarr 2016). Nonetheless, in 2013 alone, the Russian budget under-received almost \$9.5 bn because of various special provisions for oil and gas exports to Belarus. To be objective, as Russia obtained a sort of compensation of about \$3.8 bn in the form of cheaper Belarusian oil products re-exported to Russia, the net oil and gas transfer from Russia to Belarus amounted to some \$5.7 bn in 2013 (Kirkham 2016). In 2015, Russia's oil subsidies to the Belarussian economy were estimated at \$4 bn (Ravsky 2016). Lately, due to significantly lower oil prices, Belarus' margin has been on the decline resulting in Russia's subsidisation being under \$1 bn in 2016 (Ravsky 2016). In 2015, Russia was to supply 23 mn tons of oil to Belarus, and Belarus was to supply 1.8 mn tons of petroleum products to Russia (Tarr 2016), but with the oil price falling in 2015 (reducing its subsidy), Belarus refused to supply the produced petrochemicals to Russia.

Some estimates suggest that in 2015 Russian oil and gas subsidies to Belarus were still corresponding to around 15 – 18 per cent of Belarusian GDP (Roberts and Moshes 2016). Similarly, economies of Kyrgyzstan and Armenia are benefiting from stable, sufficient and affordably priced access to energy resources and this constitutes one of the points of their interest to the EAEU.

Significant disproportion in intra-trade, a result of economic asymmetry accompanied by structural homogeneity addressed above, additionally hinders EAEU integration. Trade with third countries accounts for 85.8 per cent of the entire EAEU's turnover, with Russia holding the highest share of such trade - 91.3 per cent, Kazakhstan - 78.1 per cent, Armenia - 70.8 per cent and Kyrgyzstan - 64.1 per cent. In contrast, the intra-trade is of more importance to Belarus, as 52.2 per cent of its turnover is placed within the EAEU. Considering the total export of the EAEU, Russia dominates with 62.4 per cent while Belarus and Kazakhstan add 26.5 per cent and 9.2 per cent, respectively. In the total import of the EAEU, Belarus yet again leads with the highest among the member-states share of 36.6 per cent, followed by contributions by Russia and Kazakhstan, of 34.1 per cent and 23 per cent, respectively.

Additionally, the problem of EAEU integration can be identified in relation to the ability of the national governments and the EAEU as a whole to finance their respective national programmes of modernisation and structural transformations, as well as joint integrative undertakings. In 2016, the government's share in capital investment was beyond 24 per cent in Belarus, around 16 per cent in Kazakhstan and Russia and near 9 per cent in Armenia and Kyrgyzstan. The most productive type of investment, in R&D, remains unacceptably low. In 2016, the R&D financing amounted to 1 per cent to GDP in Russia, some 0.2 per cent in Kazakhstan and less than 0.1 per cent in Armenia and Kyrgyzstan. Given the absolute size of the national GDP even of the largest EAEU economies, it is hard to speak of meaningful investment in innovations. On the other hand, private investments in the EAEU economies are discouraged by a host of factors. Particularly for Russia, the problem has been not a lack of national private capital, but a poor business environment. Even though capital outflow from

Russia has lessened, according to Russia's Central Bank, from $152.1 bn in 2014 to $58.2 in 2015 and further down to $19.2 in 2016, this is sooner a result of the limitations imposed by the sanctions rather than due to Russia's improved investment climate. Moreover, the sanctions caused a dramatic decline in foreign direct investment to Russian economy: from $44.1bn in 2014 to $13bn in 2015.

While EAEU economies' own financial resources are seriously limited, their banking sector remains heavily state-controlled and the foreign-owned banks' activity is severely regulated. The share of foreign banks in the total number of banks operating in EAEU as a whole is 18.5 per cent, in Russia – 13.8 per cent (in 2016 the number of banks controlled by foreign capital declined by 1/4 as compared to 2013), in Kazakhstan – 42.4 per cent (the number of foreign banks decreased by 1/4), in Kyrgyzstan – 44 per cent, in Armenia – 47.4 per cent and in Belarus 58.3 per cent (the number of foreign-controlled entities fell by nearly 1/3).

Overall, the asymmetries in the size of economies, in the distribution of internal and external trade, in the gains from external tariff structure and other aspects effectively render the supranational regulatory governance having correspondingly asymmetrical influence. The EAEU's founding treaty envisions that all decisions at the highest level must be reached by consensus under "one country, one vote" rule. Nonetheless, as Russia is geographically, demographically and economically substantially larger and it funds almost 80 per cent of the EAEU integration initiatives (Gusev & Westphal 2015), it is rather plausible to presume that Russia's self-interests, as well as the situation in the Russian economy essentially determine the pattern of the EAEU's integration and the dynamism thereof (van der Togt et al. 2015).

4. The Common Gas Market of the EAEU

The beginning of the joint energy initiatives was announced on November 18, 2011, when Belarus, Kazakhstan and Russia endorsed the Declaration on Eurasian Economic Integration. By doing so, they confirmed the member-states intentions towards the creation and effective functioning of the common energy market, implementation of coordinated energy policy and harmonisation of energy legislation. The EAEU's common energy market initiative implies the subsequent creation of the Common Electricity Market (by 2019), the launch of the Common Market for Oil and Oil Products (by 2020) and the establishment of the Common Gas Market (by 2025). Formation of each of the markets is a gradual process, which is to be coordinated by the Minister of Energy and Infrastructure and the respective divisions within the departments of Energy and Transport and Infrastructure. Respective regulatory institutions have been formed subsequently (Figure 4).

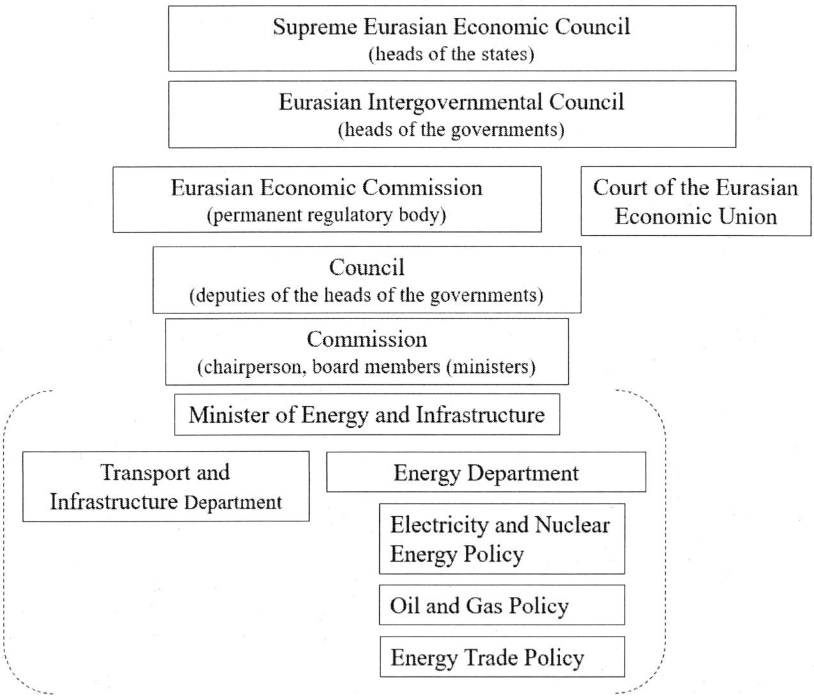

Source: Author.

Figure 4. Institutions for Energy Integration within EAEU

The Energy Department of the Eurasian Economic Commission functions in collaboration with respective national government entities on three competences: economy, energy and antitrust regulation and competition. Interestingly, Kazakhstan and especially Belarus have slightly different approach on the latter. Competition promoting (monopoly preventing) agency is missing in Belarus, while Kazakhstan's antitrust agency simultaneously oversees the matters of natural monopolies.

4.1. Institutionalisation of the CGM

On May 31 2016, the Supreme Eurasian Economic Council adopted the Concept for the Common Gas Market of the Eurasian Economic Union (Concept). The Concept defines CGM as "a system of trade and economic relationships among the economic agents of the member states in relation to the transportation and supply of natural gas among the member states".[35] The CGM is sought to enable trade in natural gas among the EAEU member states; and transportation of natural gas owned by the member states and designated for the delivery to the member states. CGM is projected as yet another indispensable mechanism to promote economic exchange among the EAEU member states, thereby enhancing full-fledged economic integration within the EAEU.[36]

The Concept states that the formation of the CGM shall be pursued while taking into consideration: the priority of the member states' individual international commitments before respective third parties; the specifics of functioning and transformation of the national gas markets in the member states; the supremacy of satisfying the domestic gas demand over the demand of EAEU members; the extant nation states' regulatory provisions for their domestic gas markets; and the international practices as they relate to the common gas markets' formation.

[35] Author's translation of a respective definition provided in Concept (2016).

[36] Para 1 Article 79, Para 2 Article 83, para 4 and 5 of Article 104 and Appendix 22 of the Treaty (2014).

According to the Concept, the CGM's goals are to: enhance energy security of member-states; improve reliability, accessibility and quality of gas supply; and increase the efficiency of use of the member-states' national gas transporting systems. The CGM's objectives include: provision of non-discriminatory access of all members to the CGM; enhancement of fair competition among the market participants (not limited to gas suppliers); gradual transition to market pricing; active use of national currencies in gas trade; promotion of favourable investment climate for the common gas transportation system (GTS) development; ensuring transparent access to the GTS capacities; harmonisation of norms and standards in GTS; and exchange of information about the pricing, wholesale prices and transport tariffs in national gas markets.

The realisation of the CGM is planned in three stages. Till 2020, there are activities towards regulatory, technical and technological harmonisation, information exchange and development of other instruments necessary for the CGM functioning, such as the rules of access to the GTS and the codes for trade exchange, the development of the indicative aggregate EAEU's gas balance and others. By the end of 2021, the trade exchanges shall start operating; the volume of long-term contracting shall be defined so that the GTS' expansion would be implemented accordingly, including the augmentation of the GTS' capacities through the joint investment schemes; the coordination of competing interests of member-states' gas suppliers in respective markets shall be completed; among other targets and others. By January 1, 2025, the CGM shall be fully operating supported by free trade in gas via direct contracts and gas trade exchange schemes and market pricing ensuring equal profitability across the CGM.

The Concept names respective national gas regulators, as well as designated CGM's regulatory agencies, as responsible for the regulatory governance as they regard natural monopolies, anti-trust regulation, tariff regulation, technical regiments and regulation of gas trading at exchanges.

Based upon the provisions of the Concept, the principal model of the CGM can be depicted as follows (Figure 5).

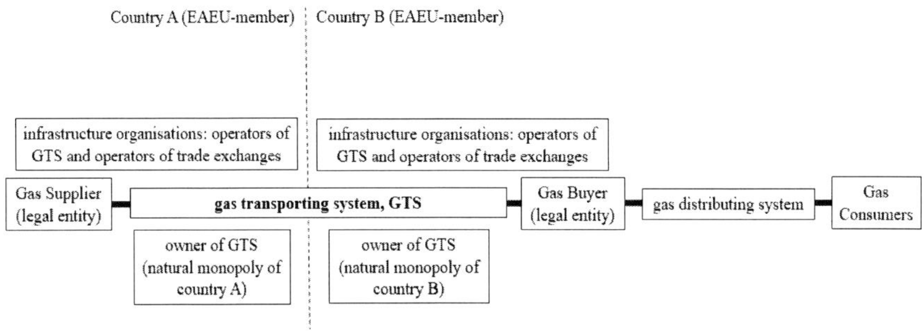

Source: Author, based on the Concept (2016).

Figure 5. Principal Model of CGM.

The Concept outlines several forms and instruments for the CGM's regulation (Figure 6).

Types of Activity

- Transportation
- Delivery/ supply
- Direct trade (via contracts)
- Trade at commodity exchange
- Unification and harmonisation of standards for gas transmission systems

Gas Market Regulation

- Regulation of natural monopolies
- Antitrust regulation
- Customs and tariff regulation
- Technical reglamentation
- Exchange trade regulation

Regulatory Instruments

- Regulatory provisions for the CGM
- Development of the CGM indicative gas balance
- Harmonisation of regulatory environments in member states
- Analysis of CGM development
- Development and application of common rules of access to gas transmission system
- Collaboration among national infrastructure organisations

Source: Author, based on the Concept (2016).

Figure 6. Forms and Instruments of the CGM Regulation.

The pricing mechanism described in the Concept proposes a voluntary commitment of the member-states to introduce market-based gas price. By 1 January 2025, a decision is to be reached on defining gas price in the EAEU on a netback basis, meaning net of the costs associated with bringing gas to a respective market. The Concept does not contain provisions for gas trade outside the EAEU; it is implied that respective bilateral agreements will continue to govern gas trade with non-EAEU countries. As a preparatory step, the EAEU members will examine whether there is any conflict of interest in relation to export to third-party markets.

Apparently, the creation of a full-fledged competitive gas market is not an ultimate objective of the EAEU's CGM initiative. The proposed institutional design demonstrates consolidation of member-states in their acceptance of large-scale monopolistic actors backed by respective national governments with a limited inclusion of competition-enhancing instruments, such, for instance, as trading via gas exchanges.

4.2. The EAEU's Economies Gas Markets

The EAEU economies are very divergent in terms of their gas profiles (Table 2), which in this case is a positive trait inducing an essential for integration complementarity. There is a giant gas producer and exporter of the world scale - Russia, as well as small consumer markets absolutely lacking their own gas resources - Armenia and Kyrgyzstan. The EAEU economies import over 73 per cent of gas from within the EAEU with Russia being the major supplier. Only Kazakhstan imports some natural gas directly (not via Gazprom) from Uzbekistan and Turkmenistan. In 2016, these imports were 1.5 bcm and 1.1 bcm, respectively.

Table 2. EAEU Member-States' Gas Profiles, 2015 (bcm)

	Armenia	Belarus	Kazakhstan	Kyrgyzstan	Russia	EAEU
Production	-	0.2	45.5	0.03	633.4	679.3
Consumption	2.5**	19.5	16.4*	0.4**	444	~482.8
Export, total	-	-	12.7	-	192.4	205.1
Export to third countries	-	-	3.9	-	169.1	173
Export within EAEU	-	-	8.8	-	23.3	32.1
Import, total	2.3	18.8	5.8	0.4	22.3	49.6
Import from third countries	0.4	-	3.2	-	9.7	13.3

Notes: (*) Kazenergy 2015; (**) CIA Factbook (2014).

Source: Author based on Gazprom and EAEU statistics.

Table 3. Connectivity to Gas Pipeline Network in EAEU Member-States, 2016

	Armenia	Belarus	Kazakhstan	Kyrgyzstan	Russia
Share of population with access to pipeline gas, %	95	99	uneven: 79.9-96.4 in the west; 19.1-63.4 – in the south; the north (except for Kostanay region) and the east have no access	22	67.2 (highly uneven, eastern regions have largely no access)

Source: Author based on Gazprom and EAEU statistics.

Some clarifications are due regarding Kazakhstan's gas statistics. The BP data differ tremendously from that provided by the EAEU and Kazakhstan's KazMunayGaz itself. In fact, Kazakhstan produces over 45 bcm of natural gas annually, but only less than 20 per cent of this volume is being commercially utilised/domestically sold (this is what the BP statistics reflects), the rest is used for reinjection at the oil fields in order to increase oil output. Kazakhstan relies on coal for over 60 per cent, while oil and gas occupy almost 20 per cent and 17 per cent in the country energy mix, respectively. Nearly 98 per cent of gas reserves are located in the west and this fact shapes the national pipeline network in a highly uneven manner. On average, slightly over 40 per cent of the population has access to pipeline gas in Kazakhstan. Yet, Kyrgyzstan is the least interconnected among the EAEU gas markets (Table 3). Relatively low access ratio for Russia is due to the peculiarities of its geographical, demographic and economic profiles: vast eastern territories with very low density of population and limited economic activity are largely not embraced by the pipeline network.

Having no natural gas of its own, Belarus has nonetheless been in a privileged position in terms of import prices owing to its special relations with Russia. Russian-Belarusian gas relations are induced by mutual dependency. Indeed, Russia is a major gas supplier to Belarus. However, owning to the fact that Russia is a significant exporter to the vast European markets, Belarus possesses a leverage over Russia for the transit of 32.5 bcm Europe-bound gas via the Yamal-Europe pipeline traversing Belarusian territory. Mutual dependency results in Belarus having a special (discounted) import price, while Russia has a special (discounted) transit tariff. To be precise, Belarus has been paying price equivalent to that for Russia's domestic customers in Yamal-Nenets Okrug (the region of the origin of Russian gas) plus transport tariff incurred to deliver gas via the trunkline to Belarus (Manenok 2016, Ravsky 2016). Such provisions resulted in Belarus' import prices being two-fold lower than the average Russian prices for Germany and 2.5 times less than those for Ukraine.[37]

With detailed data for the intra-EAEU gas export during the entire period of 2011-2016 being unavailable, Figure 7 demonstrates that trade in fossil fuels constitutes major share of Russia's export in the EAEU.

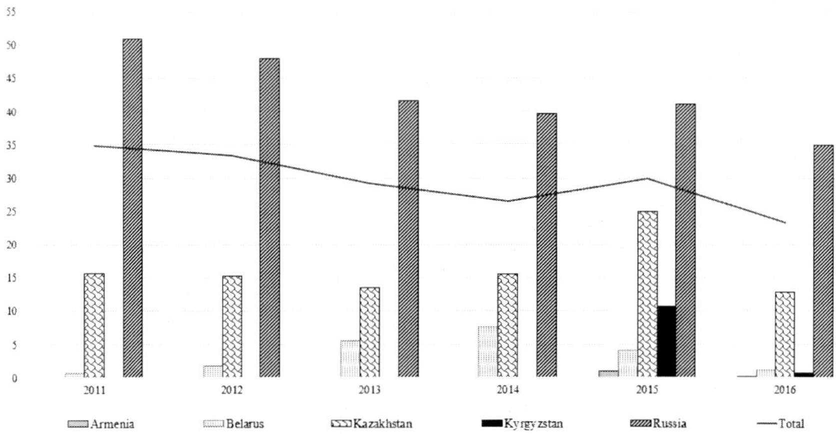

Source: Author, based on http://eec.eaeunion.org/ru/act/integr_i_makroec/dep_stat/tradestat/tables/Pages/default.aspx.

Figure 7. Export of Mineral Fuels, Oil and Oil Products (HS 27) within Customs Union/ EAEU (% to total).

[37] By the same token of mutual dependency, Ukraine was also having special prices prior to 2014. This was in exchange for the benefits that Russia received in the form of the lease of the naval bases at the Black Sea, discounted transit tariffs and many other instruments, which the parties regularly (re)negotiated over. Following the Crimean crisis, the setting has drastically changed, but as Ukraine is not included in the scope of this study, such analysis is omitted.

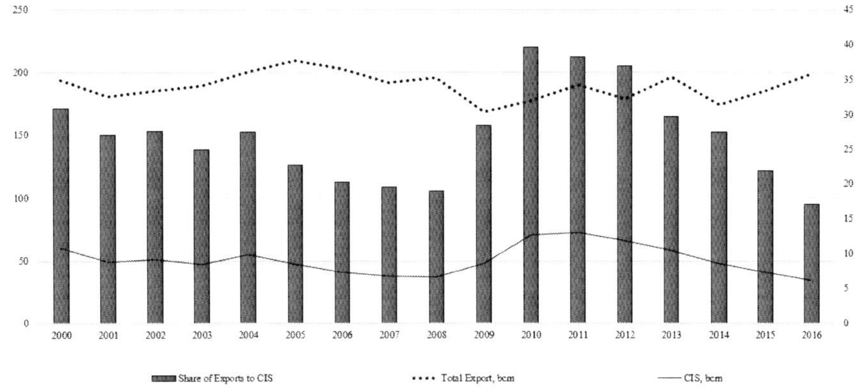

Source: Author, based on CBR data.

Figure 8. Russia's Natural Gas Export to CIS (bcm and %, rhs).

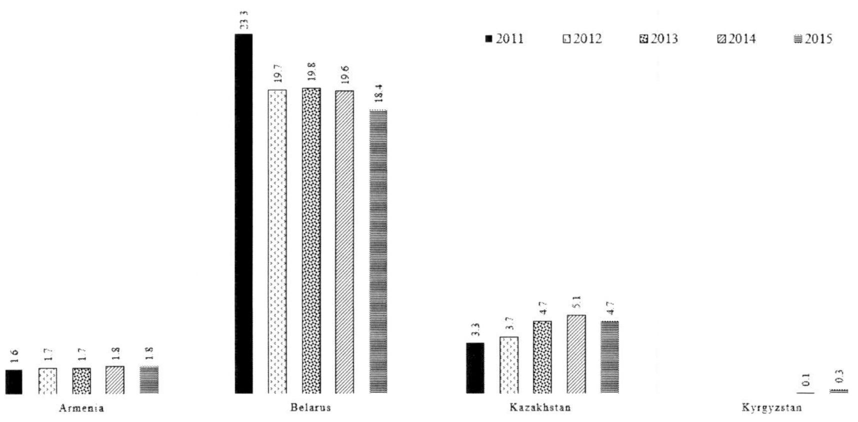

Source: Author, based on Gazprom data.

Figure 9. Russia's Natural Gas Export to EAEU (bcm).

Trade in natural gas between resource-rich Russia and other EAEU members represents relatively small fraction of Russia's respective exports. The European markets have traditionally been and remain of key importance to Russia.[38] Moreover, Figure 8 demonstrates that while gas

[38] In 2016, Russia exported 214 bcm of gas, including record high 164 bcm (13.8 per cent increase against the year 2015) to the non-CIS markets. The largest increases were accounted for Germany and the UK (for more detail see: "Statistika vneshnego sektora"

supplies to third countries has been on a rise, Russia's exports to the CIS markets (to which the CGM members all belong) have been declining in both volume and share representation.

Gas export from Russia to Belarus is much higher than in the case of other EAEU states (Figure 9). The two neighbours are closely linked by the cross-border value chains, as demonstrated earlier. 90 per cent of electricity supply for the Belarusian industry is generated from Russian gas. Additionally, Belarusian refineries process almost exclusively Russian crude oil with the products being re-exported mainly to Russia. Under the common energy markets framework, the two countries' already closely linked businesses are likely to deepen their ties further.

At the same time, Russia's gas giant Gazprom continues to import natural gas from Central Asia. In general, the company practices two types of transactions: trading and swap contracts.

In 2015 Gazprom imported 22.3 bcm of natural gas, of which the largest share was bought in Kazakhstan (12.6 bcm), followed by Uzbekistan (3.5 bcm) and Turkmenistan (3.1 bcm). In 2016, Gazprom purchased 6.2 bcm in Uzbekistan and 12.9 bcm in Kazakhstan.[39] Initially, the principal reason for such a practice was Gazprom's attempt to fulfil its obligations before its customers in the so-called Far Abroad (non-CIS economies) and up until to 2014 – Ukraine. At the time of high energy prices, Gazprom has been benefiting from arbitrating: buying at low prices (under the intergovernmental agreements) in Central Asia and re-selling at higher prices in the European markets. Gazprom was additionally benefitting from so called re-export regime, which effectively exempted the company from paying Russia's export duties. This was criticised by Gazprom's main contender – Rosneft – Russia's largest state-owned oil company which aggressively increases natural gas output and lobbies for the liberalisation of gas export.

<https://www.cbr.ru/statistics/?PrtId=svs> and "TEK Rossii 2016" (in Russian) (accessed June 1, 2017).

[39] When not specified, data are collected from various accounts released by Gazprom.

Overall, as European demand for Russian gas has become less certain (due to various factors, such as the EU's climate change policy, economic downturn, political tensions with Russia over the Ukrainian situation and others), as well as Russia's domestic demand has been stagnating, Gazprom, whose production capacity steadily exceeds its export obligations (Shadrina 2017a), has naturally become disinterested to maintain long-term import contracts with the Central Asian suppliers. Currently, Gazprom emphasises its interest to develop cooperation with those Central Asian countries which let the company operate in their upstream sectors. Such is the case of Uzbekistan, from which Gazprom intends to increase imports by 4 bcm from the year 2018 in return for the equity in Uzbekistani projects among other things. On the contrary, following the price dispute with Turkmengas, Gazprom terminated its importing obligations in 2016 and only recently announced a possibility to resume transactions provided the sides manage to settle the dispute and Turkmenistan (similarly to Uzbekistan) offers broader business opportunities to Gazprom in its domestic sector.[40]

Gazprom applies swap and cross-border trading with Kazakhstan, Uzbekistan and Kyrgyzstan. On swaps, Kazakhstan supplies Russian Orenburg Gas Processing Plant with gas from its Karachagan filed, while Gazprom, in turn, supplies Aktyubinsk and Kostanai regions of Kazakhstan with Russian gas. On cross-border trade, Gazprom purchases gas from Uzbekneftegaz to supply southern regions of Kazakhstan, in particular, Almaty region or Kyrgyzstan. By numbers, to supply southern Kazakhstan, in 2015 Gazprom purchased 2.9 bcm in Uzbekistan, which was 20 per cent less than year before. Similarly, in 2015 Gazprom purchased 0.02 bcm and 0.2 bcm of gas from Uzbekistan and Kazakhstan, respectively, to supply Kyrgyzstan. Such schemes are entirely explained by the respective gas companies' attempts to overcome the deficiencies of their domestic transport pipeline sectors and optimise their costs while exploiting the inherited from the Soviet era cross-border transport systems.

[40] Petlevoi, Vitaly (2017) Gazprom kupil uzbekskii gaz. *Vedomosti*. 6 aprelya (in Russian).

Certainly, the CGM is a small market whose capacity stands no comparison with the European import potential. Yet, having faced with the changing institutional environment in the EU (the European Third Energy Package,[41] the Central and South-Eastern Europe Gas Connectivity project[42], etc.) and increased gas-to-gas competition, Gazprom on the one hand has markedly adjusted its business strategy in the European markets, but it also became more attentive towards export strategies of other potential suppliers, such as Azerbaijan and Turkmenistan. For its massive export potential, Turkmenistan's ambitions expanding towards Gazprom's traditional markets in Europe is of special concern to Gazprom. Additionally, Turkmenistan's already firm position in the Chinese market makes Gazprom more inclined to resume negotiations with Turkmenistan although imports under deliberation are considerably smaller and being conditioned by Gazprom's access to Turkmenistan's gas sector. The Chinese market has all along been attractive to Gazprom. After having eventually commenced its ever-expensive pipeline project called the Power of Siberia, the company put itself in a rather insecure position,[43] and it now realises that increasing competition in the European and Chinese markets requires a more nuanced export strategy.

To summarise, three principal considerations can be extended to explain Gazprom's motivation in the CGM: first, an attempt to diversify its exports (even if the volumes are insignificant), especially given that the infrastructure is already in place and no major investment is required; second, an attempt to expand its transactions via its own subsidiary companies and affiliated businesses in the CGM countries; and third, an attempt to improve its prospects for possible new joint export schemes in cooperation with producers in Central Asia (Kazakhstan, Uzbekistan, Turkmenistan) and Caucasus (Azerbaijan).

---

[41] Third Energy Package <http://www2.nationalgrid.com/UK/Industry-information/Europe/Third-energy-package/> (accessed September 25, 2017).

[42] Central and South Eastern Europe Gas Connectivityconnectivity> (accessed September 25, 2017).

[43] *Pravda. July 4, 2017* <https://www.pravda.ru/economics/materials/gas/04-07-2017/1340650-siberia_china-0/> (accessed September 25, 2017).

Overall, Russia's interest in the CGM can be interpreted as an attempt to reinforce economic links with the EAEU member states. The CGM implying the formation of a net of long-term joint institutions is one of such effective means to reach Russia's goals for Eurasian integration. The importance of gas is uneven for the EAEU economies ranging from high (for Belarus and Armenia) to low (Kazakhstan and especially Kyrgyzstan), but gas-importing economies stand to derive substantial economic benefits from the CGM. The latter is especially applicable to Belarus, which even attempted to influence the sequence of energy integration, insisting that CGM needs to be launched before the common electricity market (Pastukhova & Westphal 2016).

5. Discussion: Institutionalisation of CGM

This section intends to project theoretical perspectives of institutionalism presented in the opening section on the studied case by discussing CGM institutionalisation via two particular notions, those of institutional change and institutional complementarity.

5.1. Complementarity of Institutional Environment within EAEU

As has been presented, the EAEU's institutionalisation has eventually embraced the features of supranationalism and multilateralism the best role model of which is the EU. Unlike in the EU members, the institutional environment in the EAEU economies is distinguished by the endorsement of the norms of state capitalism. Indeed, the share of the state in national economies as of 2015 was estimated at 70 per cent of GDP in Russia (35 per cent in 2005), 70-75 per cent in Belarus and about 60 per cent in

Kazakhstan.[44] We particularly emphasise this dimension of an overall institutional environment in the EAEU – prominent role of the state in economy – as a critically important generic measure. The latter allows us to characterise institutional environment in EAEU members as considerably homogeneous. We observe substantial cohesiveness of the national institutional settings across various sectors and co-ordinated evolutionary transformations in national institutional systems (Hall and Soskice 2001, Rustemova 2011). We narrow our discussion to the case of CGM to provide more nuanced arguments.

5.2. Evolutionary Institutional Complementarity within CGM

In the following, we demonstrate that a result of incremental institutionalisation of EAEU, CGM utilises the virtues of institutional compatibility and reveals the logic of institutional conversion among the national markets of the member-states. We scrutinise institutional changes in the CGM by exploring the organisational principles and the structure of CGM, as well as trading in CGM, paying utmost attention to gas pricing and price.

5.2.1. Organisational Structure of CGM

Table 4 provides comparative overview of the national gas markets in the EAEU. None of five national markets can be characterised as competitive. In such a context, the Russian gas industry even appears to be slightly more advanced for its (yet rather careful) application of exchange trade and competition-enhancing regulation (letting non-Gazprom actors engage into production and other segments of the domestic value chain). Apparently, Russian Gazprom has business presence in all EAEU markets, but most extensively it is involved in the gas sectors of Armenia and Kyrgyzstan.

---

[44] Evraziiskaya ekonomicheskaya integratsiya – 2017. Evraziiskii bank razvitiya. Doklad № 43, 2017, s. 68 (in Russian) <http://eurasian-studies.org/wp-content/uploads/2017/05/EDB_Centre_2017_Report_43_EEI_RUS.compressed.pdf> (accessed July 25, 2017).

Table 4. Characteristics of National Gas Markets

Segment of Gas Industry	Armenia	Belarus	Kazakhstan	Kyrgyzstan	Russia
Output	no domestic production	single producer Belarusneft	some features of monopolistic structure; larger presence of foreign capital	single actor Gazprom Kyrgyzstan (subsidiary of Russia's Gazprom)	decreasing influence of monopolist Gazprom
Trunklines, transmission and transit systems	single owner and operator so-called natural monopoly Gazprom Armenia (subsidiary of Russia's Gazprom), which is gas supplier and owner of transport system	owner, operator and single supplier Gazprom Transgaz Belarus (subsidiary of Russia's Gazprom)	single owner and operator natural monopoly KazTransGas (subsidiary of KazMunayGas), transport and supply bundled	single owner and operator natural monopoly Gazprom Kyrgyzstan (subsidiary of Russia's Gazprom); transport and supply bundled	four owners and operators, natural monopoly: Gazprom, Noril'skgazprom, Yakutgazprom and Rosneft'-Sakhalinmorneftegaz; transport and supply bundled
Storage	single owner and operator Gazprom Armenia	single owner and operator Gazprom Transgaz Belarus	single owner and operator KazTransGas	does not exist	segment of transport system; non-natural monopoly; single owner and operator Gazprom
Distribution	single owner and operator Gazprom Armenia	single owner and operator Beltopgaz	single owner and operator KazTransGas	single owner and operator Gazprom Kyrgyzstan	partly competitive segment; Gazprom dominates
Wholesale supply and retailing	supply monopoly; total price regulation and control; no exchange trade	supply monopoly; total price regulation and control; no exchange trade	total price regulation and control; no exchange trade	supply monopoly; total price regulation and control; no exchange trade	competitive market; multiple suppliers; price regulation; exchange trade since October 2014
Operative monitoring and control over transport system	by the owner of transport system	by the owner of transport system	by the owner of transport system	by the owner of transport system	by the owners of transport system

Source: Author, based on respective companies' data.

Overall, it appears that institutional setting of Russian gas sector is projected to the entire CGM. This may well be due to the fact that Russia is the key supplier to the markets in question as well as because Russian state-owned Gazprom through its subsidiaries has dominant positions in all the member-states of the EAEU except Kazakhstan (where such role is played by the national champion KazMunayGas). In Belarus Gazprom is involved in the onward shipping of Russian natural gas via the Yamal-Europe pipeline, through its subsidiary Transgas. In Kyrgyzstan, Gazprom possesses exclusive rights to explore and exploit the domestic gas fields until 2028 and has taken over the state-owned Kyrgysgas. In the Armenian gas sector, Gazprom possesses a monopoly via a subsidiary. In Kazakhstan, Gazprom operates a joint venture (KazRosGas) and runs joint gas and oil exploration projects with KazMunayGas as well as with other Russian companies (Rosneft and Lukoil).

Despite the fact that Gazprom's financial leeway has become considerably weaker at this time of lower energy prices and its reputational status in the European markets has been undermined by a long-practiced politicisation of gas relations with the EU, affiliated with Russian Gazprom businesses dominate in the EAEU (except for Kazakhstan where national champions operate in a similar to Gazprom mode). Previously dominated by Gazprom - a textbook case of state-owned company with an exclusive right to export all Russia-origin gas, Russia's gas sector is no longer the case of absolute monopoly: in 2016, Gazprom produced 64.4 per cent of Russia's gas, which is significantly less than its share of 84.7 per cent in 2006. In addition to the strengthening positions of Rosneft and Novatek, there is a sizable progress in the output by so-called independent gas producers, whose number grew to 159 and their share in total gas output reached almost 14 per cent, as of 2016.[45] Gradual demonopolization in production is, nevertheless, not matched by the liberalisation of pipeline gas export, which remains the prerogative of Gazprom. Moreover, Gazprom is a sole owner of Russia's gas pipeline system, which effectively blocks access to the transport system for non-Gazprom producers unless

[45] TEK Rossii 2016, s. 36.

they have an agreement with the monopolist. Yet, there is certain anticipation that persistent lobbying by Rosneft (Russia's third largest gas producer with 46.7 bcm of output, as of 2016) will result in some format of liberalisation in pipeline export (analogous to the adopted in 2009 liberalisation of LNG export).

The CGM is informed by two suppliers (Russia and Kazakhstan, more specifically, respective national champions) and three consumers (Armenia, Belarus and Kyrgyzstan, often affiliated with Russia's monopoly Gazprom). This setting makes two types of negative externalities probable: a lack of competition between the suppliers; and the probability of the bargaining power exercised by consumers. Whether Russia and Kazakhstan[46] will be cooperating or competing depends largely on gas price. If the price in the CGM is comparable to that in the respective national markets of Russia and Kazakhstan, but significantly lower than in the external export markets (such as the EU or China), the prospects for competition between Russian and Kazakhstani suppliers will be undermined by the possibility of the suppliers' preference for non-CGM markets. Certainly, such deviating behaviour can be corrected via a kind of supply quota set directly by a supranational regulatory authority, which will effectively debilitate market nature of the CGM.

There is a probability for third parties' exports to the CGM. Currently, Iran is supplying Armenia with some gas (about 0.4 bcm in 2016) swapping it for the electricity. Apparently, Iran is keen to increase its gas export to Armenia. Provided the upgrades in the pipeline capacity are made, it can supply up to 3 bcm annually.[47] In such a context, a feasible market strategy for Russia to ensure its future place in Armenia's small market is to continue to maintain a competitive price. Russia's gas relations with Georgia, which is a transit country for Russian gas transported to Armenia, constitute an additional dimension of the ties within the CGM. Currently, Gazprom pays in-kind for the transit,

---

[46] Kazakhstan has recently announced its plans to export 7 bcm annually to China.

[47] Iransky gaz mozhet zamenitj dlya Armenii rossiisky, no on gorazdo dorozje <http://www.soyuzinfo.am/rus/analitics/detail.php?ELEMENT_ID=440> (accessed May 20, 2017).

supplying Georgia with one tenth of the export volumes passing through the Georgian territory. From 2018, however, the transit fee will be determined in monetary terms.

Discussing the matter of third parties' role in the CGM is incomplete without examining the role of China. So far, Russia has not been particularly enthusiastic about the materialisation of the prospects where it would need to share its gas export ambitions for the Chinese market with other Central Asian suppliers. However, Russia's Gazprom dominance in the Central Asia was discontinued after China completed the construction of a large-scale gas pipeline from Turkmenistan in 2009. Possible construction of a second pipeline, the Line D, through Uzbekistan, Tajikistan and Kyrgyzstan necessitates Russia's closer cooperation with China in the region. Gas in particular and energy cooperation on the whole can be part of a broader cooperation agenda. As Xin (2016) and Ziguo (2016) argue, aligning the Silk Road Economic Belt (SREB) and the EAEU in terms of hard infrastructure would help create the Eurasian economic space.

Although the CGM is currently envisioned as a non-competitive market, we allow for a possibility of a scenario that the progression with the liberalisation of the Russian gas market (over varied reasons, domestic or/ and international) would trigger relevant changes in the institutional provisions of the CGM. In other words, institutional arrangements for gas market in member states will convert towards those altered in the Russian market so that institutional complementarity among national gas markets could be regained and the CGM remained functioning.

5.2.2. Gas Trade in CGM

Considering gas price and pricing in the CGM, while the free market pricing is not the case for any of the five national markets, the mode of price regulation and its principles are varying (Table 5). In Russia, gas price is regulated by the Federal Tariff Agency and monitored by the Anti-trust Agency. So far, relatively small-scale trade in gas has been practiced at the Saint Petersburg International Mercantile Exchange. In Belarus, gas price is determined by the government. The CGM Concept envisages that pricing will be implemented in a way enabling the coordination of the interests of all CGM members. Although the initial organisation of the CGM appears to serve mostly to the protection of the currently practiced setting of natural monopoly, the Concept envisages the establishment of the international mercantile exchange for gas trading within the CGM and the unbundling in the gas sector.

Table 5. EAEU Member-States' Domestic Gas Prices and Market Characteristics, 2016

	Armenia	Belarus	Kazakhstan	Kyrgyzstan	Russia
Gazprom's export price, $/1,000 cm (as of January 1, 2017)	150	141.1	n/a	150	-
Domestic price, $US/1,000 cm households commercial sector	320 280	112 276	47 50	223 270	83 67
Export duties, %	-	-	-	-	30
Transport tariff, $US/1,000 cm/ km	20.2	1.3	3.7	1.6	0.8
Share of gas trade via commodity exchange, %	-	-	-	-	0.12

Source: Author based on Gazprom and EAEU statistics.

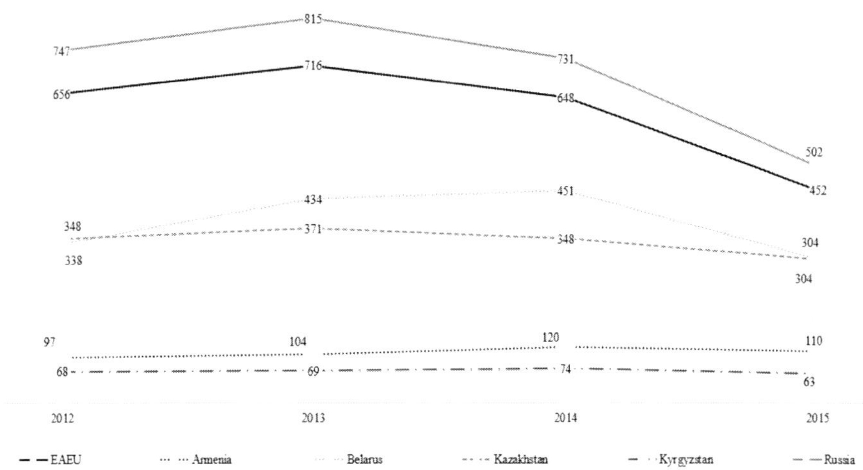

Source: Author based on data http://eec.eaeunion.org/ru/act/integr_i_makroec/dep_stat/econstat/Pages/statpub.aspx.

Figure 10. Monthly Per Capita Income in EAEU Economies, 2011-2016 (US$).

Given rather divergent levels of income in the CGM economies (Figure 10), gas price and pricing is a socially sensitive matter for all the EAEU economies and one of the most complex agendas to settle.

Kazakhstan supplies national consumers at the cheapest among other CGM members price. Among the importers, the highest gas prices are in Armenia, while the lowest is in Belarus. In Armenia and Russia, gas price for domestic commercial sector is lower than for the households, while the rest of the group practice the reverse. The largest differential between the price for the residents and the industrial users is in Belarus - almost 60 per cent, whereas in Kyrgyzstan it is about 17 per cent and in Kazakhstan almost 6 per cent.

The proposed approach to pricing in the CGM is based upon the consideration that an equal profitability across the national markets shall be assured. This, in turn, involves the notion of bargaining power. So far, as exemplified earlier, Belarus has been enjoying this leverage in its gas relations with Russia, but continuation of such practice appears to be impossible in the context of CGM. Even though Belarus is by far a significant consumer among the CGM members and as a transit state it indeed possesses the power to influence Russia's market position in the EU, once the CGM is established, Russia will no longer have a right to award Belarus the terms more favourable than those for Armenia or Kyrgyzstan. Restructuring, or rather eliminating, the existing system of the bilateral gas agreements among the CGM members and creating a transparent and uniform pricing mechanism is one of the most challenging reforms to implement. We tend to place this kind of transformation under the logic of earlier described layering (adding new elements). Once the transparent pricing is achieved, it will signify Russia's explicit departure from the realm of politicisation of its gas trade, which will have positive impact on Russia's gas relations with the EU.

Conclusion

Summing up our analysis of institutionalisation of the CGM in the context of institutional evolution of the EAEU, we conclude that the emergence of the CGM, in Aoki's (2001: 333) expression, is a purely intended fit (as opposed to drift type changes). As has been presented earlier, the institutional setting for the CGM is being designed deliberately so that it would allow the member-states to utilise their already practiced national regulatory provisions without challenging the compatibility of the principal elements of their national regulatory environments. The sequencing of the CGM's launch follows the logic of Aghion & Blanchard (1994) and Roland (2000) that the changes that are expected to be more popular are implemented first. Additionally, given social significance of this commodity, faster transition of the national gas markets characterised by a great price diversity, towards a common market with a unified free-market price converged at a higher level, would raise demand for social policies to compensate for the transitional losses, increasing thereby the costs of the CGM initiative.

Although the EAEU's CGM is in its nascent form, its publicised provisions inform a vision that the initial design of the CGM builds upon the extant institutional complementarity informed by a dominant role of state in the national economies in the member-states. More specifically, the CGM fully replicates institutions currently effective in Russia, which has the largest gas market and most diversified gas sector among the EAEU members.

The organisational structure of the CGM is projected upon the existing modes of the national gas markets: respective state-owned companies, natural monopolies in their character, are envisioned to remain the principal actors. Given that Russia's Gazprom already has its subsidiaries in EAEU economies, it appears that the initial organisational changes will be insignificant. Some market transformations, such as larger trading via gas exchanges, full-fledged unbundling of gas production, transition, distribution and sale, and others, will be introduced gradually. It seems plausible that CGM evolution will follow the transition trends in Russia's gas sector.

Gas price and pricing systems appear to be among the most complex issues stretching beyond the EAEU boundaries. This is the realm where Russia is likely to incur what many assess as Russia's forgone benefits effectuated by the integration. However, taken in a broader context, the integration of gas markets in the EAEU implies the discontinuation of the practice of price multiplicity, which will help Russia mend its soured reputation vis-à-vis the European consumers; the benefits of such recovery might be worth the costs.

FUNDING ACKNOWLEDGMENT

This research was supported by a grant-in-aid from the Joint Usage and Research Centre Program by the Institute of Economic Research, Hitotsubashi University (2017), JSPS KAKENHI Grant Number JP16KT0044 and Waseda University grant-in-aid 2017S-158.

REFERENCES

Aghnion, Philippe and Olivier J. Blanchard. 1994. "On the Speed of Transition in Central Europe". *NBER Macro Annals.*

Amable, Bruno. 2000. "Institutional Complementarity and Diversity of Social Systems of Innovation and Production. " *Review of International Political Economy.* 7(4):645-687.

Amable, Bruno. 2003. *The Diversity of Modern Capitalism.* Oxford University Press.

Amable, Bruno. 2016. "Institutional Complementarities in the Dynamic Comparative Analysis of Capitalism." *Journal of Institutional Economics.* 12(1): 79-103.

Andrews-Speed, Philip. 2016. "Applying institutional theory to the low-carbon energy transition." *Energy Research & Social Science.* 13: 216–225.

Aoki, Masahiko. 1994. "The Contingent Governance of Teams: Analysis of Institutional Complementarity." *International Economic Review.* 35(5):657-676.

Aoki, Masahiko. 2000. *Information, Corporate Governance, and Institutional Diversity*. Oxford University Press.

Aoki, Masahiko. 2001. *Toward a Comparative Institutional Analysis.* Cambridge: MIT Press.

Aoki, Masahiko. 2007. "Endogenizing institutions and institutional changes." *Journal of Institutional Economics,* 3 (1): 1–31

Aslund, Anders. 2013. "Push for Customs Union Turns Friends to Foes." *The Moscow Times*. September 10.

Crouch, Colin. 2010. "Complementarity" in Glen Morgan, John L. Cambell, Colin Crouch, Ove Kaj Pedersen and Richard Whitley (eds.) *The Oxford Handbook of Comparative Institutional Analysis*. Oxford University Press.

Tarr, David G. 2016. The Eurasian Economic Union of Russia, Belarus, Kazakhstan, Armenia, and the Kyrgyz Republic: Can It Succeed Where Its Predecessor Failed? *Eastern European Economics*, 54 (1): 1-22.

Deeg, Richard. 2005. "Path Dependency, Institutional Complementarity, and Change in National Business Systems." In *Changing Capitalisms? Internationalisation, Institutional Change and Systems of Economic Organisation,* edited by G. Morgan, R. Whitley and E. Moen. Oxford University Press

Deeg, Richard. 2007. "Complementarity and Institutional Change, In Capitalist Systems," *Journal of European Public Policy.* 14(4): 611-630.

Dragneva, Rilka and Kataryna Wolczuk. 2012. Russia, the Customs Union and the EU: Cooperation, Stagnation or Rivalry? *Briefing paper REP BP 2012/01*, Chatham House.

Dragneva, Rilka and Kataryna Wolczuk. 2013. The Eurasian Customs union: Framing the analysis. In *Eurasian economic integration: Law policy and politics,* edited by R. Dragneva & K. Wolczuk. Edward Elgar Publishing.

Dragneva, Rilka and Kataryna Wolczuk. 2017. "The Eurasian Economic Union Deals, Rules and the Exercise of Power." *Chatam House Research Paper.*

Dreyer, Iana and Nicu Popescu. 2014. "The Eurasian Customs Union: The Economics and the Politics." *European Union Institute for Security Studies. Brief No. 11.*

Gusev, Alexander and Kirsten Westphal. 2015. "Russian Energy Policies Revisited. Assessing the Impact of the Crisis in Ukraine on Russian Energy Policies and Specifying the Implications for German and EU Energy Policies." German Institute for International and Security Affairs. *SWP Research Paper* #8.

Hall, Peter A. 1999. The Political Economy of Europe in an Era of Interdependence" in *Continuity and Change in Contemporary Capitalism,* edited by Herbert Kitschelt, Peter Lange, Gary Marks, John D. Stephens. Cambridge University Press.

Hall, Peter A. and Daniel W. Gingerich. 2009. "Varieties of Capitalism and Institutional Complementarities in the Political Economy: An Empirical Analysis." *British Journal of Political Science.* 39 (3): 449-482.

Hall, Peter A. and David Soskice. 2001. "An Introduction to Varieties of Capitalism", in *Varieties of Capitalism: The Institutional Foundations of Comparative Advantage*, edited by Peter A. Hall and David Soskice. Oxford: Oxford University Press.

Hartwell, Christopher A. 2013. "A Eurasian (or a Soviet) Union? Consequences of further economic integration in the Commonwealth of Independent States." *Business Horizons.* 56, 411-420.

Hasanov, Fakhri, Cihan Bulut and Elchin Suleymanov. 2017. "Review of energy-growth nexus: A panel analysis for ten Eurasian oil exporting countries." *Renewable and Sustainable Energy Reviews.* 73: 369-386.

Hodgson, Geoffrey. 2006. "What Are Institutions?" *Journal of Economic Issues* 40 (1): 1-25.

Höpner, Martin. 2005. "What connects industrial relations and corporate governance? Explaining institutional complementarity." *Socio-Economic Review*, 3(2): 331–358.

Kazenergy. 2015. *The National Energy Report*. Almaty. 2015. http://www.kazenergy.com/images/NationalReport15_English.pdf.

Kirkham, Ksenia. 2016. "The formation of the Eurasian Economic Union: How successful is the Russian regional hegemony?" *Journal of Eurasian Studies.* 7: 111-128.

Kleffner, Jan K. 2003. The Impact of Complementarity on National Implementation of Substantive International Criminal Law". *Journal of International Criminal Law. 1(1): 86-113.*

Laure Delcour (2015). Between the Eastern Partnership and Eurasian Integration: Explaining Post-Soviet Countries' Engagement in (Competing) Region-Building Projects. *Problems of Post-Communism*, 62 (6): 316-327.

Lochwood, Watthew, Caroline Kuzemko, Catherine Mitchell and Richard Hoggett. 2017. "Historial institutionalism and the politics of sustainable energy transitions: A research agenda." *Environment and Planning C: Politics and Space.* 35(2): 312-333.

Makocki, Michal and Nicu Popescu. 2016. China and Russia: an Eastern partnership in the making? *Chaillot Paper No. 140. European Union Institute for Security Studies.*

Mostafa, Golam. 2013. "The concept of 'Eurasia': Kazakhstan's Eurasian policy and its implications." *Journal of Eurasian Studies.* 4: 160-170.

North, Douglass C. 1990. *Institutions, Institutional Change, and Economic Performance*. Cambridge: Cambridge University Press.

North, Douglass C. 1991. "Institutions." *Journal of Economic Perspectives* 5 (1): 97–112.

North, Douglass C. 1994. "Economic Performance through Time." *American Economic Review* 84(3): 359–67.

North, Douglass C. 1995. “Five Propositions about Institutional Change.” In *Explaining Social Institutions*, edited by Jack Knight and Itai Sened. Ann Arbor, Mich.: University of Michigan Press: 15–26.

North, Douglass C. 1997. “Prologue.” In *The Frontiers of the New Institutional Economics*, edited by John N. Drobak and John V. C. Nye. San Diego and London: Academic Press: 3–28.

Parkhomchik, Lidiya. 2016. *Natural Gas Industry of Kazakhstan: Key Features and Future Prospects*. <http://eurasian-research.org/en/research/comments/energy/natural-gas-industry-kazakhstan-key-features-and-future-prospects> (accessed May 30, 2017).

Pastukhova, Maria and Kirsten Westphal. 2016. “A Common Energy Market in the Eurasian Economic Union Implications for the European Union and Energy Relations with Russia”. *German Institute for International and Security Affairs. SWP Comments.*

Popescu, Nicu. 2014. “Eurasian Union: The Real, the Imaginary and the Likely.” *Chaillot Paper, no. 132, European Union Institute for Security Studies.*

Roland, Gérard. 2000. *Transition and Economics: Politics, Markets, and Firms (Comparative Institutional Analysis).* MIT University Press.

Rustemova, Assel. 2011. “Political economy of Central Asia: Initial reflections on the need for a new approach.” *Journal of Eurasian Studies*. 2: 30-39.

Sean P. Roberts & Arkady Moshes. 2016. The Eurasian Economic Union: a case of reproductive integration? *Post-Soviet Affairs*, 32 (6): 542-565.

Shadrina, Elena. 2016. “External Factors for Russia’s Economic Growth. An Overview of Relevant Theoretical and Empirical Studies.” *Journal of Siberian and Far Eastern Studies*. 14: 30-71.

Shadrina, Elena. 2017a. “Russia’s Oil and Gas for Northeast Asian Markets: Means and Ways vs Realities.” *Journal of Self-Governance and Management Economics*. 5(1): 34–79.

Shadrina, Elena. 2017b. “A Comparative Check on Inclusiveness of Economic Growth and Development: Russia vis-à-vis some Post-

Soviet Economies." *Journal of Siberian and Far Eastern Studies*. 16: 8-45.

Shumylo-Tapiola, Olga. 2012. The Eurasian Customs Union: Friend or Foe of the EU? Carnegie Europe: Carnegie Endowment for International Peace.

Svarin, David. 2016. "The construction of 'geopolitical spaces' in Russian foreign policy discourse before and after the Ukraine crisis." *Journal of Eurasian Studies*. 7: 129-140.

The Eurasian Economic Union: Power, Politics and Trade. 2016. *Europe and Central Asia Report N°240*. International Crisis Group.

van der Togt, Tony, Francesco Saverio Montesano and Iaroslav Kozak. 2015. From Competition to Compatibility. Striking a Eurasian balance in EU-Russia relations. *Clingendael Report*. Netherlands Institute of International Relations.

Vsevolod Samokhvalov. 2016. The new Eurasia: post-Soviet space between Russia, Europe and China. *European Politics and Society*, 17 (sup1): 82-96.

Xin, Li. 2016. *Chinese Perspective on the Creation of a Eurasian Economic Space*. Moscow. Valdai Club. <http://valdaiclub.com/files/12585/> (accessed May 25, 2017).

Ziguo, Li. 2016. *Eurasian Economic Union: Achievements, Problems and Prospects*. China Institute for International Studies. <http://www.ciis.org.cn/english/2016-08/19/content_8975486.htm> (accessed May 25, 2017).

(in Russian)

Concept. 2016. *"O кontseptsii formirovaniya obschego rynka gaza Evraziiskogo ekonomichskogo soyuza."* Vysshii Evraziiskii ekonomicheskii sovet. Astana. Reshenie № 7. 31 maya 2016. [On *concept for Common Gas Market for EAEU*. Supreme Eurasian Economic Council. Astana. Directive No.7. May 31, 2016].

Manenok, Tatiana. 2016. *Pochemy Belarusi segodnya slozhnee torgovatjsya s "Gazpromom"?* [*Why is it more difficult for Belarus to negotiate with Gazprom?*] <http://www.belrynok.by/2016/01/15/pochemu-belarusi-segodnya-slozhnee-torgovatsya-s-gazpromom/> (accessed June 1, 2017).

Ravsky, Maksim. 2016. *"Gazovy gesheft"* - vsyo: ischezaet odna iz opor sovremennoi belorusskoi ekonomiki. [*Gas lucre* – finished: one of the pillars of modern Belarusian economy is to disappear] <http://bdg.by/news/authors/gazovyy-gesheft-vsyo-ischezaet-odna-iz-opor-sovremennoy-belorusskoy-ekonomiki> (accessed June 1, 2017).

TEK Rossii 2016. 2017. Analitichesky Tsentr pri Praviteljstve Rossiiskoi Federatsii. [Fuel energy complex of Russia. Analytical Centre for the Government of the Russian Federation].

Treaty. 2014. *"Dogovor o Evraziiskom ekonomicheskom siyuze"* Vysshii Evraziiskii ekonomicheskii sovet. Astana. 29 maya 2014 [*The Treaty on the Eurasian Economic Union.* Supreme Eurasian Economic Council. Astana. May 29, 2014].

About the Editor

Dr. Sreekanth. K. J, Associate research Scientist at Kuwait Institute for Scientific Research (KISR) is specializing in Energy policy planning and energy efficiency studies. Presently he is working in the Energy efficiency technologies program under Energy and Building Research Center (EBRC) at KISR. After completing his post graduation, he joined in Ashok Leyland Ltd, Chennai, India, as a Senior Development Engineer in R & D Department in 1999. Later he joined in Mar Baselios College of Engineering and Technology, University of Kerala, India as a lecturer and worked there until 2014. He held various positions under the University, including the Head of Department of Mechanical Engineering. He has more than 25 international journal & conference publications as well as book chapters. His PhD research was mainly concentrated on the estimation of CO2 emission reduction and development of a methodology for energy policy planning for the domestic energy sector in Kerala, a state of India. However, the methodology arises out of thE work can be extended not only to any region of the country but also to any part of the world. Presently his research area is Energy Efficiency studies and Energy Policy Planning, which is, developing innovative solutions to reduce per capita energy consumption by applying energy-efficient technologies to different sectors such as buildings, road transportation and to develop advanced solutions for applications of renewable energy technologies such as biofuels, suitable for the climate. The major aim in these studies is to reduce per capita primary energy consumption without affecting the quality of life while adhering to environmental standards.

INDEX

#

A

B

C

D

E

F

G

H

I

Q

R

S

T

U

V

W

Y